Mathematics

8A

Sue Gunningham
Pat Lilburn

OXFORD
UNIVERSITY PRESS
AUSTRALIA & NEW ZEALAND

OXFORD
UNIVERSITY PRESS
AUSTRALIA & NEW ZEALAND

Level 8, 737 Bourke Street, Docklands, Victoria 3008, Australia

Oxford University Press is a department of the University of Oxford. It furthers the University's objective of excellence in research, scholarship, and education by publishing worldwide in

Oxford New York

Auckland Cape Town Dar es Salaam Hong Kong Karachi Kuala Lumpur Madrid Melbourne Mexico City Nairobi New Delhi Shanghai Taipei Toronto

With offices in

Argentina Austria Brazil Chile Czech Republic France Greece Guatemala Hungary Italy Japan Poland Portugal Singapore South Korea Switzerland Thailand Turkey Ukraine Vietnam

OXFORD is a trademark of Oxford University Press in the UK and in certain other countries

First published 2009
Reprinted 2009, 2010 (twice), 2013 (twice), 2014 (twice), 2015, 2018, 2022, 2023, 2025

ISBN 978 0 19 556250 7

Typeset by Sunset Digital Pty Ltd, Brisbane
Illustrated by Uramina and Nelson Pty Ltd and Guy Holt
Printed in China by Golden Cup Printing Co. Ltd

Oxford University Press Australia & New Zealand is committed to sourcing paper responsibly.

Every effort has been made to trace the original source of copyright material contained in this book. The publisher would be pleased to hear from copyright holders to rectify any errors or omissions.

Table of Contents

For Students

Dear Student,

We hope you enjoy working through the Oxford Mathematics books for Grade 8. The two books have been written to show how mathematics is useful in dealing with the everyday world. The lessons will help you to gain important life skills that will be useful now and in the future when you leave school to work for someone else or to set up your own small business.

The books comprise four topics:

Book A **Topic 1:** Gadgets

Topic 2: And the Winner is . . .

Book B **Topic 3:** A Lot Like Me

Topic 4: The Tourism Market

Each topic is broken into three *Learning Units* that will build on your previous knowledge and introduce you to new concepts. Each Learning Unit begins with pictures and open-ended questions to explore your knowledge of both the context and some of the mathematics to be covered.

Help Boxes have been located throughout the Learning Units to provide clear instructions to support your understanding of particular concepts. We encourage you to refer back to these Help Boxes as you progress through the book to refresh your memory and consolidate your understanding. A number of *Challenges* also appear in the Learning Units. It is hoped that you will attempt as many of these Challenges as possible to extend your thinking.

A Revision Unit, titled *Additional Learning, Revision and Assessment*, appears at the end of each topic after the three Learning Units. This unit has four parts. It contains one stand-alone lesson with a specific maths focus that is separate from those concepts contained in the three Learning Units. The stand-alone lesson is followed by a revision exercise that allows you to revise your understanding of the work covered in the topic. The unit also provides one investigative group task and a formal test.

A *Glossary* appears at the back of each book. The words in the glossary relate specifically to the content of that particular book and the information provided should make it easier for you to understand the meaning of different mathematical terms.

An answer section has been included at the back of each book.

We hope you find the books an interesting and challenging part of your learning journey.

The Authors

Topic 1 Gadgets

Learning Unit 1: Mobile Phones

Learning Unit 2: Wheels

Learning Unit 3: In the Tool Shed

Learning Unit 4: Additional Learning, Revision and Assessment

Topic 1 Gadgets

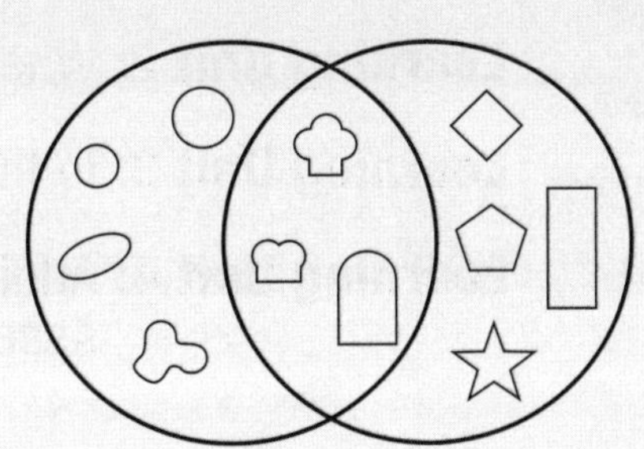

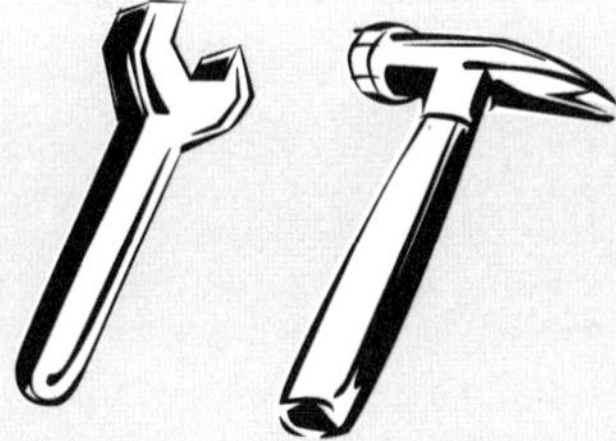

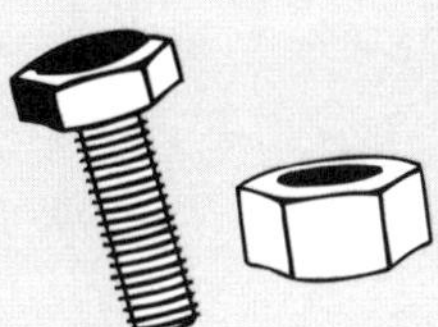

In this topic you will explore the mathematics associated with a range of gadgets and tools found in the home and beyond.

The topic includes a study of wheels and the special features of a number of measurement tools commonly found in a tool shed. It also contains important information about some of the fees and charges associated with the use of mobile phones.

The Assessment Tasks provide opportunity for you to demonstrate your understanding of the material covered through both project work and a formal test.

Topic 1: *Gadgets* comprises the following Learning Units:

Learning Unit 1: Mobile Phones

Learning Unit 2: Wheels

Learning Unit 3: In the Tool Shed

Learning Unit 4: Additional Learning, Revision and Assessment

Overview

Each Learning Unit reflects an aspect of mathematics used in everyday life. A brief summary of the mathematics in each Learning Unit appears below.

In **Learning Unit 1**, *Mobile phones*, the work deals mainly with using time and money rates to calculate telephone charges. The unit also involves solving problems by factorising and simplifying algebraic equations.

Learning Unit 2, *Wheels*, focuses on the properties of circles, and applying the relevant formulae to work out the diameter, circumference and area of different sized circles. The unit also includes some work on Venn diagrams.

The work in **Learning Unit 3**, *In the Tool Shed*, involves working with angles and identifying the size of unknown angles based on the size of known angles within the same shape. The unit also investigates the use of fractions and decimals to size different tools.

Learning Unit 4, *Additional Learning, Revision and Assessment*, includes a stand-alone lesson about directed numbers, a revision lesson, an investigation and a formal test.

Some of the materials that you may need to complete the activities in this topic are listed below.

You may need:

rulers

protractor

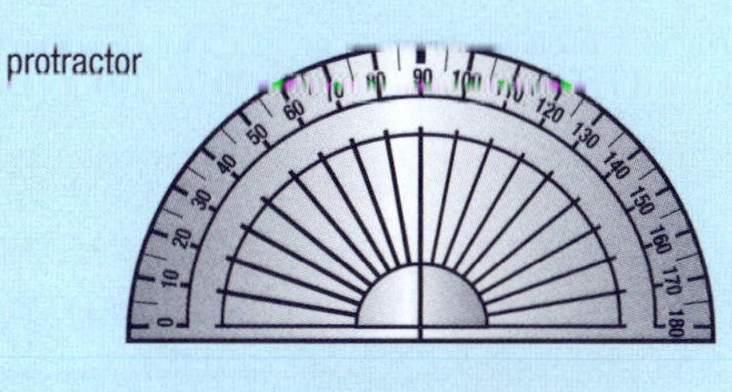

compass

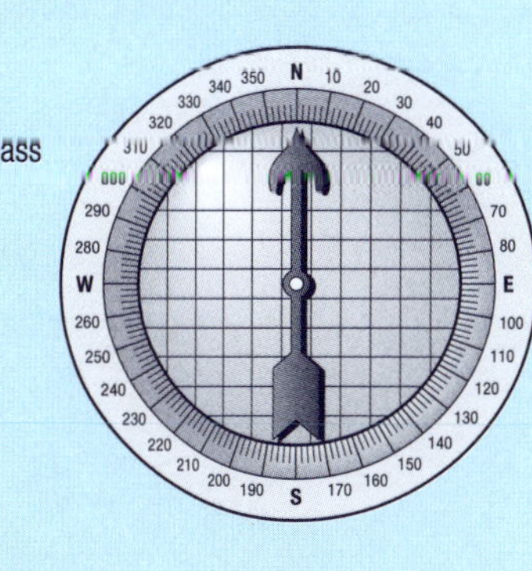

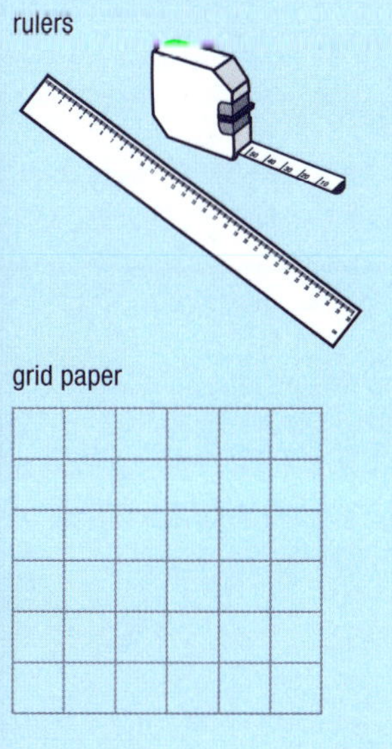

grid paper

calculator

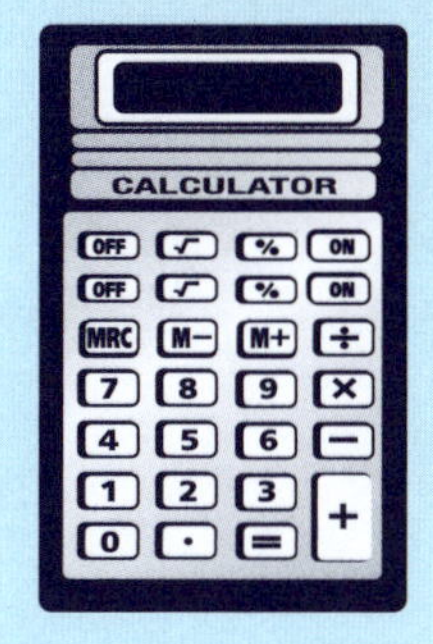

paper and pencil for drawing

Learning Unit 1 Mobile Phones

Strand: Number and Application

Ratios and Rates	Outcome 8.1.6	Apply rates to solve simple problems from real life
Directed Numbers	Outcome 8.1.7	Apply directed numbers in problem solving

Strand: Measurement

Time	Outcome 8.3.5	Use time–rate calculations

Strand: Patterns and Algebra

Algebra	Outcome 8.5.3	Manipulate simple algebraic expressions and solve real life problems

Lesson 1:	**Introduction**	Mobile phones payment options—prepaid, pay-as-you-go, fixed contract Advantages and disadvantages of each option
Lesson 2:	**Mobile phone charge rates**	Charge rates Comparing different rates Calculating costs and times based on charge rates TXT rates
Lesson 3:	**Using algebra to calculate costs**	Applying substitution to calculate call costs
Lesson 4:	**Some of this and some of that**	Simplifying algebraic expressions using addition and subtraction
Lesson 5:	**What did it cost me?**	Simplifying simple algebraic expressions Using simple algebra to solve problems
Lesson 6:	**More time, more money**	Using the distributive law to calculate the cost of multiple time-blocks Using the distributive law to expand algebraic expressions
Lesson 7:	**Keep it simple**	Writing simple algebraic expressions to describe situations Expanding and simplifying algebraic expressions
Lesson 8:	**Substitution**	Evaluating algebraic expressions
Lesson 9:	**Factorisation**	Finding the highest common factor in a set of numbers Using factors to solve simple equations
Lesson 10:	**Mobile phone debt**	Adding and subtracting directed numbers involving money.

Lesson 1 Introduction

During this unit you will revise and extend some of the work that you did in Grade 7 about time and money. You will use your understanding of numbers and patterns to explore mobile phone plans and calculate charges.

The sudden increase in mobile phone use has created a communication revolution around the world. It is therefore important to be aware of the costs involved in owning a mobile phone and what to look for when signing a mobile phone contract. The three main ways that people pay for their mobile phone calls are:

- prepaid phone calls
- pay-as-you-go contracts
- fixed term contracts

1 How many people do you know who own a mobile phone?

2 What are some advantages of owning a mobile phone?

3 List some disadvantages of owning a mobile phone.

4 What are some of the additional features that are included with many mobile phones?

The three different ways that people pay for their mobile phones are described in the table below. Each option has advantages and disadvantages. People should think seriously about their own phone needs and the costs involved before deciding the best option for them.

Prepaid	Pay-as-you-go	Fixed term contract
Supply your own phone. You pay an amount (e.g. K30) and this amount is coded into your phone so that you can use the phone until this amount is used up. You pay again to top up the account.	Usually supply your own phone. You get a bill each month for the number of calls made. There is no contract so you are not locked into a phone plan for a set time at a set rate.	You sign with one company for a specific period of time (e.g. 36 months) with fees and a specific charge for calls. You usually get a new phone as part of the deal.

5 Which plan do you think is likely to be the most popular with teenagers? Why?

6 Which plan do you think is the least expensive? Why?

7 Which plan do you think is the most expensive?

8 What might you need to think about before deciding which plan best suits you?

Lesson 2 Mobile phone charge rates

Mobile phone companies usually charge a set rate for each 30-second block of time that people speak on their phone. The charge rate per 30-second block varies from company to company.

Help Box

In Grade 7 you learned that 'rate' is a comparison between quantities of different kinds.

The most commonly used rate is 'speed' where the distance travelled is compared to the time taken.

For example, a person who cycles 24 kilometres in 2 hours is cycling at the average rate of *12 kilometres per hour* (or 12 kph).

$$\text{Speed} = \frac{\text{Distance travelled}}{\text{Time taken}}$$

Mobile phone charge rates compare call costs with the number of 30-second blocks taken on the phone.

For example, a person charged 87 toea for speaking on their mobile phone for 3 (30-second) blocks is being charged at the rate of *29 toea per (30-second) block.*

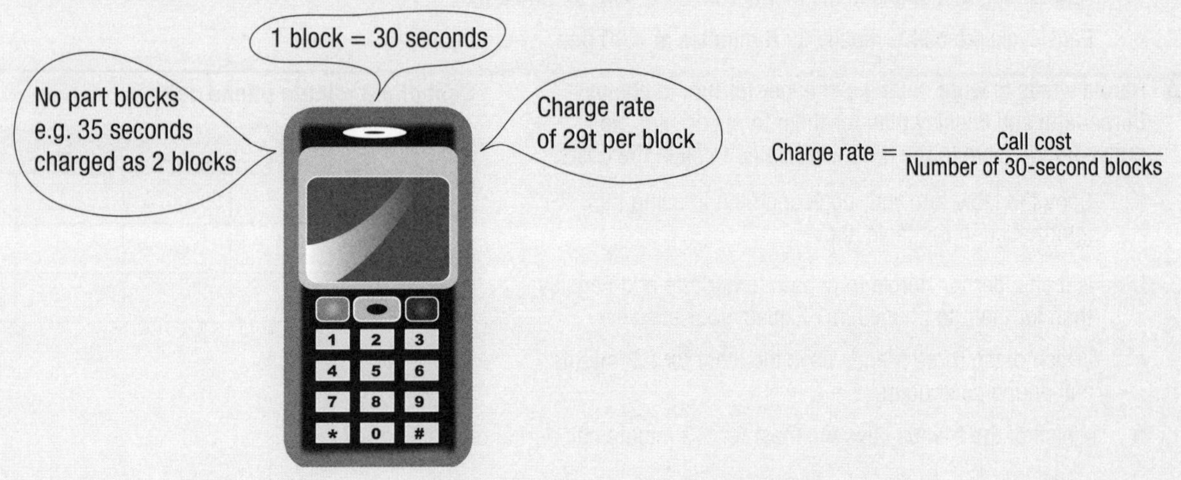

$$\text{Charge rate} = \frac{\text{Call cost}}{\text{Number of 30-second blocks}}$$

1 Express each of these as a charge rate assuming the company uses a 30-second time block.

- **a** K1 for a 1 minute call
- **b** 90 toea for a $1\frac{1}{2}$ minute call
- **c** K1.60 for a 4 minute call
- **d** 96 toea for a 90 second call
- **e** K2.50 for a $2\frac{1}{2}$ minute call
- **f** K1.20 for a 5 minute call

2 Calculate the cost of these calls if the charge rate is 29 toea per 30-second block.

- **a** 2 minute call
- **b** 90 second call
- **c** $4\frac{1}{2}$ minute call
- **d** 35 second call
- **e** 47 second call
- **f** 2 min 15 second call

3 Calculate the minimum time a person must have talked on their mobile if they were charged the following amounts at a charge rate of 33 toea per 30-second block.

- **a** 99 toea
- **b** K2.31
- **c** K1.32
- **d** K1.98
- **e** K3.30
- **f** K3.63

Phone companies often charge different rates depending on what time of day a call is made. The table below shows the difference between call costs charged by three mobile phone companies.

	Company A used by Bernadette	Company B used by Harold	Company C used by Fegsley
Peak time charges 9 a.m.–6 p.m.	50 toea per 30-second block	40 toea per 30-second block	1 toea per second, same cost applies 24 hours a day
Off-peak charges 6 p.m.–9 a.m.	15 toea per 30-second block	20 toea per 30-second block	

4 Use the information in the table to calculate the cost of the following calls:

- **a** Bernadette phoned a friend and spoke for 1 minute at 11 a.m.
- **b** Bernadette phoned a friend at 9 p.m. and spoke for 4 minutes.
- **c** Harold spoke on his mobile for 5 minutes at 7.30 a.m.
- **d** Harold rang and spoke to his friend from 8.55 a.m. to 9.05 a.m.
- **e** Fegsley spoke on his mobile for 8 minutes at 4.30 p.m.

Challenge

Bernadette spoke on her mobile phone from 5.54 p.m. for 20 minutes. What did the call cost her?

5 Harold wants to work out if it's cheaper for him to phone Bernadette and Fegsley than for them to phone him. He draws up the table to the right and begins to fill in the costs.

- **a** Copy the table into your book and fill it in using the information in the table above.
- **b** Is it cheaper for Harold to phone Bernadette and Fegsley than for them to phone him? Explain your answer.
- **c** Which of the three friends pays the least for a 3 minute call during peak hours?
- **d** Which of the friends pays the most for a 3 minute call during off-peak time?

Cost of a 3 minute phone call

	8 a.m.	noon	6 p.m.	10 p.m.
Bernadette				
Harold				
Fegsley				

6 Bernadette's brother asks if he can use her mobile phone to call some friends. He gives her K5 to cover the costs. Use the rates for Bernadette's phone in the table in question 5 to answer the following questions.

- a How many seconds can Bernadette's brother talk during peak time for K5?
- b How many seconds can he talk during off-peak time for K5?
- c What length of time could he talk if he used Harold's phone instead and paid him K5? Give the answers for peak time and off-peak time.
- d How long could he talk if he paid Fegsley K5 and used his phone instead? Give the answers for peak time and off-peak time.

The local newspaper advertised a free phone as part of a 24-month contract.

Free Phone

K0

on a K40 plan
for 24 months

Call charges 25t per 30 sec block.
Minimum charge K40 per month

7 What is the maximum time I could talk on a phone each month at the charge rate in the advertisement if I wanted to make sure that I did not have to pay more than K40 per month?

8 Maria speaks for an average of 15 minutes a day on her mobile phone.

- a What will be her phone call charge at the end of 4 weeks if currently she is being charged at the rate of 30 toea per 30-second block?
- b Would it be financially sensible for Maria to sign up for the phone contract in the advertisement? Include an explanation of costs in your answer.

It is usually cheaper to TXT a message using a mobile phone than to make a phone call. Most TXT charges are based on a 160-character block. A 'character' is any letter, number, space or symbol.

TALK'S CHEAP . . .

TXT
9t for up to
160 characters

Call charge
15t per 20 sec block

9 Use the information in the advertisement to the right to answer the following questions.

- a What will it cost at this rate to send a TXT of 45 characters?
- b What will it cost at this rate to make a 2 minute phone call?
- c If there was only K3 credit remaining on an account being charged at this rate, how long could I talk on the phone?
- d If I was charged 18 toea for sending a TXT at this rate, what is the minimum number of characters that I could have sent?
- e If I was charged K1.05 for a phone call, what is the longest timed call that I could have made based on this rate?

Challenge

Many people use abbreviations to reduce the number of characters used in a TXT to save time and money. For example, 'I will see you later tonight' can be reduced to 'C U L8R 2nite'.

Simplify a favourite poem into TXT. Calculate the cost of sending the complete poem compared to using the simplified TXT version based on a rate of 9 toea per 160-character block.

Lesson 3 Using algebra to calculate costs

In mathematics, letters are sometimes used as a substitute for numbers. These letters are called **pronumerals**. They are useful in helping us to describe patterns and solve mathematical problems more easily. Pronumerals that can change in value are called **variables**.

Help Box

If a person is charged at a rate of 10 toea per 20-second block, we multiply the number of blocks by 10 to calculate the total cost of the call.

Charge rate of 10t per block

[1 block] = 10 toea

[ ] [ ] = 20 toea

[ ] [ ] [ ] = 30 toea

x × [ ] = x × 10 toea

= 10x toea

1 The diagram below shows the length of a person's call in 30-second blocks. What is the total cost of the call if the block rate is:

- a 23 toea per block?
- b 40 toea per block?
- c 35 toea per block?
- d *w* toea per block?

[1 block] [1 block] [1 block]

2 The diagram below shows five 30-second blocks. Calculate the cost of the call if the block rate is:

- a 15 toea per block
- b 21 toea per block
- c *w* toea per block
- d *k* toea per block

[1 block] [1 block] [1 block] [1 block] [1 block]

3 The table shows the call and TXT rates charged by three different companies. Calculate the different amounts a person will be charged by each company for making a:

- a 3 block phone call
- b 10 character TXT
- c 5 block phone call
- d 300 character TXT

Company	Call rate (30 sec block)	TXT (160 characters)
Alpha Company	25 toea	13 toea
Beta Company	31 toea	9 toea
Gamma Company	29 toea	10 toea
Delta Company	*m* toea	*p toea*

Sometimes mobile phone companies add an extra charge for every call or TXT regardless of the length of the call or TXT. This extra charge is called **flagfall**.

4 A phone company charged 45 toea per 30-second block and 20 toea flagfall. Calculate the total cost of the phone call represented in the following diagram.

5 Copy the diagram for the Odyssey Mobile Phone Company into your book and write an algebraic expression to describe the cost of a call where n equals the number of 30-second blocks.

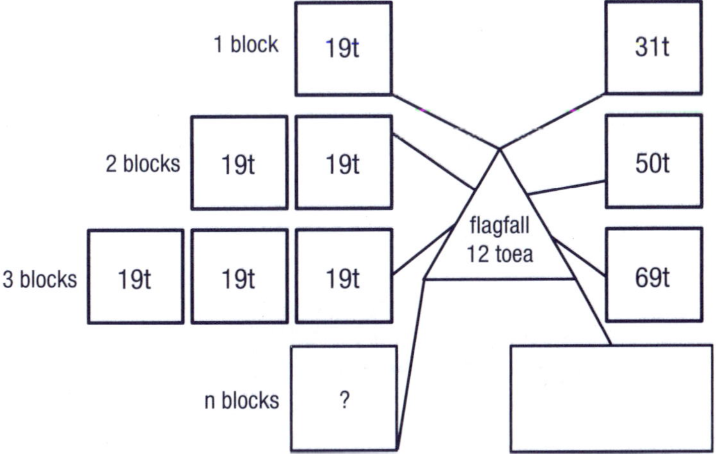

The diagram shows the block rate for calls plus the flagfall cost for the Alpha Company.

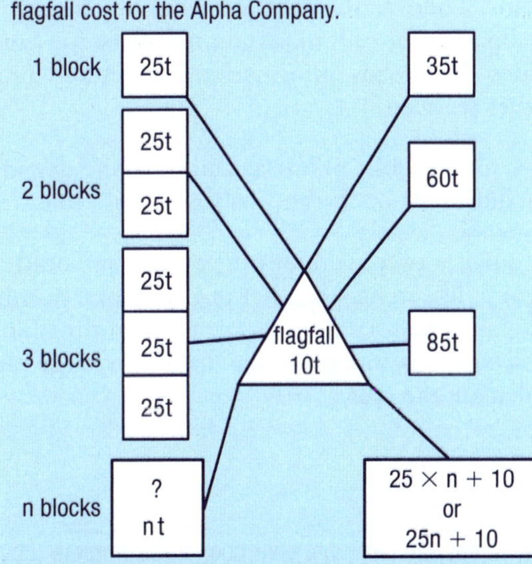

The cost of a call using the Alpha Company charge rate can be written as the algebraic expression $25 \times n + 10$ (or $25n + 10$) where n equals the number of 30-second blocks.

6 Draw a similar diagram to show the call costs for a company where the block rate is 17 toea and the flagfall is 11 toea. Show the total costs for 1, 2, 3 and n blocks.

7 Let n be the number of 30-second blocks. Use the algebraic expression $15n + 10$ to calculate the total cost of a mobile call lasting:

a 1 block **b** 4 blocks **c** 3 blocks **d** 6 blocks

8 Let n be the number of 30-second blocks. Write an algebraic expression to describe how to calculate the cost of a call where:

a the block rate is 19 toea and the flagfall is 12 toea

b the block rate is 24 toea and the flagfall is 12 toea

c the block rate is 21 toea and the flagfall is 30 toea

d the block rate is 21 toea and the flagfall is 15 toea

9 The table shows the different charge rates and flagfall for three companies. Let n be the number of 30-second blocks. Write an algebraic expression for the total cost of a phone call from each of the three companies.

10 Each of the three companies in the table also charges a flagfall on every TXT. Let z be each set of 160 characters. Write an algebraic expression for the total cost of a TXT from each of the three companies.

Company	Call rate (30-sec block)	TXT (160 characters)	Flagfall (per use)
Wan	25 toea	13 toea	10 toea
Tu	31 toea	9 toea	8 toea
Tri	29 toea	10 toea	5 toea

Lesson 4 Some of this and some of that

It can be confusing trying to work out the amounts charged for different features on some mobile phone plans. There can be special charges for ring-tones, for wallpaper, for calls to certain numbers, for music or video downloads, information services and a host of other services.

A typical mobile phone account for one day might include a list of charges such as in this table.

A quick way to work out the total cost would be to group 'like items' together (e.g. the total number of flagfalls, the total number of standard time blocks, and so on) before using the different charge rates to calculate the cost.

Phone company charges

Service	Cost per unit
1 ringtone	50 toea
1 wallpaper	K1.50 per month
3 blocks international roving	K1.80 per block
1 information call	90 toea
1 flagfall	10 toea
3 blocks standard time	25 toea per block
1 video download	K1.20 per download
1 information call	90 toea
4 blocks international roving	K1.80 per block
1 flagfall	10 toea
6 blocks standard time	25 toea per block

1 Copy the table into your book and complete it by grouping the information from the list above. The first example has been done for you.

Item	Total times listed	Charge per unit	Total charge
standard calls	$3 + 6 = 9$ blocks	25 toea per block	$9 \times 25 = 225$ toea (or K2.25)
international roving calls			
information calls			
flagfall			
wallpaper			
ringtone			
video download			

2 Choose the like terms in each of the following sets.

a $2x, 3ab, 4x$

b $k, 8ab, 2ba$

c $x, 2k, 5x$

d $3a, 2b, 2c, 2a, 2ac$

e $2mn, 3m, 3n, mn$

f $2ab, 3ac, 4abc, bc, 3abc$

Help Box

Pronumerals can be added and subtracted just like numbers. Addition can be done in any order, so $4 + 5$ has the same answer as $5 + 4$. Algebraically, in general, $a + b = b + a$.

The order is important in subtraction. For example, $5 - 4 \neq 4 - 5$. In general, $a - b \neq b - a$.

Algebraic expressions containing exactly the same pronumerals are called **like terms**. The pronumerals in like terms must all be the same but they can be in a different order. Only like terms can be added or subtracted.

For example, $5a$ and $3a$ are like terms. They can be added or subtracted, so $5a + 3a = 8a$ and $5a - 3a = 2a$.

$4mx$ and $2xm$ are also like terms. They can be added or subtracted, so $4mx + 2xm = 6mx$ and $4mx - 2xm = 2mx$.

However $3x$ and $4y$ are not like terms. They cannot be added or subtracted. $3x + 4y = 3x + 4y$

The pair $4b$ and 7 are also not like terms. They cannot be added or subtracted. $4b - 7 = 4b - 7$

3 Simplify:

a	$p + p$	b	$a + a + a + a$	c	$g + g + g$	d	$5a + a$
e	$4x + x$	f	$2y + 3y + 2y$	g	$k + 4k + 3k$	h	$7n - 3n$
i	$9m - 4m$	j	$4x - 3x$	k	$10k - 9k$	l	$a - a$

4 A phone bill listed three ringtones charged at s toea each and 14 calls charged at k toea each. Write an algebraic expression for the total cost of the ringtones and calls.

5 A person owed x kina to the phone company. She paid an amount of m kina off the bill, which meant she still owed y kina. Write an algebraic expression to describe the situation.

Challenge

Write a real-life story that could be described by the algebraic expression $3m + 2m = 5m$.

6 Simplify the following algebraic expressions by adding or subtracting the like terms.

a	$a + a + a =$	b	$5x + 2x =$	c	$7ab - 4ba =$	d	$6mn - mn =$
e	$3y + 4y - x =$	f	$4p - p + 2a =$	g	$2 + 4h + h =$	h	$7 - 4k - 2k =$
i	$6h - b + 2h =$	j	$6x + 2y - 4x =$	k	$3m + 7mn - 2m =$	l	$4xy + 3x - yx + 2x =$

7 Simplify the following algebraic expressions by adding or subtracting the like terms.

a	$8mn + 2a + 3mn + 4a =$	b	$4xy - 2x + 4xy - x =$	c	$4ab + 2a - ab + 3a =$
d	$6 + 3k - 2 + 4pk - k =$	e	$4xy + 3 + 7ab - ab + 2b =$	f	$6g + 2mn + 3p + 8 =$

8 Copy the following tables into your book and complete them.

a

+	m	$2m$	p	$2p$
m				
$3m$				
p				
$2p$				

b

+	$2x$	$x + 2$	y	$2y + 2x$
x				
$3x - y$				
y				
$y + 2$				

9 Let n be the charge per 30-second phone block.

a If Sam speaks for 3 blocks, how much will he be charged (excluding flagfall)?

b If Wendy speaks for 4 blocks, how much will she be charged (excluding flagfall)?

c What is the total cost of the two calls (excluding flagfall)?

Challenge

Write a real-life story involving mobile phone charges that could be represented by the algebraic expression $4n + 2n - p = 6n - p$.

Lesson 5 What did it cost me?

Algebra can be used to think about and solve problems, and to write about problems in a simple way.

Suppose you wanted to know the rate you had been charged per 30-second block for a mobile phone call that lasted 90 seconds.

This example could be written algebraically as:

Let c = the total call cost, then $c \div 3$ = the rate per 30-second block.

1 Based on a 3 block call with no flagfall, what is the cost per block if c equals:

a 36 b 90 c 51

d 150 e 126 f 135?

Help Box

Suppose Harold's phone bill is 17 toea more than Joseph's bill. We could say that Harold's bill is $x + 17$, where x = the amount of Joseph's bill.

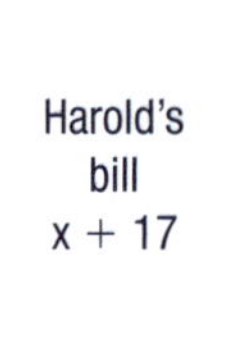

2 Let x = the amount of Joseph's bill. Write an algebraic expression to show the cost of Harold's bill if it is:

a 25 toea more than Joseph's bill

b 23 toea less than Joseph's bill

c three times as much as Joseph's bill

d half Joseph's bill

e double Joseph's bill

Help Box

Algebra uses shortcuts to make things easier. For example, usually the multiplication sign is not used.

So, instead of writing $2 \times b$, we write $2b$.

Instead of writing $3 \times b + 2 \times w$, we write $3b + 2w$.

The division sign is usually not used either.

Instead of writing $5 \div x$ we write $\frac{5}{x}$.

3 Simplify the following expressions.

a $5 \times b + 3 \times c$ b $9 \times m \div 5$

c $4 \times m + g \div 5$ d $h \div 3 - b \div 3$

e $g \times 2 \div 3 \times k$ f $c \div 3 - 2 \times a$

4 Rosa is studying her phone bills for the past few months. Suppose m is the amount Rosa had decided she can afford to pay per month for her calls.

Month	Jan.	Feb.	March	April	May	June	July
Total kina	m	$m + 7$	$m + 2$	$m - 1$	$m + 3$	$m - 4$	$\frac{m}{2}$

a Which month was Rosa's bill the highest?

b How much more was the bill that month than the amount Rosa had decided she could afford to pay monthly?

c In which months was Rosa's bill equal to or less than the amount she had decided she could afford?

d Which two-month period in the table cost Rosa the most?

5 Draw a table to show how many kina Rosa's bill would be each month from January to July if $m = 18$ kina.

6 What would Rosa's phone bill have been for each month listed if $m = 21$ kina?

7 Suppose p is the rate charged per 30-second block by the 'Phone-Home' company.

a If another company charges double that rate, how could we write this as an algebraic expression using p?

b If a different company has a charge rate of $3p$, what does this mean?

c If the 'Phone-Home' company also charges a flagfall of s, write an algebraic expression for the cost of a 30 second call.

d If the 'Phone-Home' company has a 6 month special offer and is still charging the flagfall (s) but is taking x off the cost of each 30-second block (p), write an algebraic expression for the cost of a 30 second call during the special offer.

Using the information in question 7d, write an algebraic expression for the cost of phone calls longer than one 30-second block.

Lesson 6 More time, more money

Often businesses separate out the different costs involved in the products they sell. You learned about some of these costs in Grade 7 mathematics. They include things like basic production costs, tax and profit. For example, a mobile phone company charging 25 toea per 30-second phone block may pay 15 toea in production costs and taxes and collect the remaining 10 toea as profit. This could be written as $1 \times (15 + 10) = 25$.

25 toea per time block

15t costs | 10t profit

1 Write the following as mathematical expressions using brackets.

a One block of time charged at 29 toea made up of 19 toea production cost and 10 toea profit.

b One block of time charged at 23 toea made up of 18 toea production cost and 5 toea profit.

2 The cost for 2 blocks of time charged at the same rate as the example in question 1 can be written as $2 \times (15 + 10) = 50$. Write the following as mathematical expressions using brackets and calculate the total cost of each call.

- **a** Two blocks of time charged at a rate of 19 toea production cost and 10 toea profit per block.
- **b** Two blocks of time charged at a rate of 18 toea production cost and 5 toea profit per block.
- **c** Three blocks of time charged at a rate of 15 toea production cost and 8 toea profit per block.
- **d** Five blocks of time charged at a rate of 12 toea production cost and 9 toea profit per block.
- **e** Four blocks of time charged at a rate of 15 toea production cost and 6 toea profit per block.

3 Solve these expressions by adding the numbers inside the brackets first. Then check your answers using the distributive law (see Help Box). Show your working out for both methods.

a	$5 \times (2 + 4)$	**b**	$2 \times (7 + 3)$
c	$4 \times (6 + 8)$	**d**	$9 \times (1 + 5)$
e	$3(7 + 6)$	**f**	$6(9 + 8)$
g	$10(2 + 1)$	**h**	$12(2 + 3)$

Help Box

Normally the numbers inside the brackets of a mathematical expression are worked first.

In the expression $3 \times (8 + 5)$ we work out the brackets first, then multiply the answer by 3.

$$\begin{aligned} 3 \times (8 + 5) &= 3 \times 13 \\ &= 39 \end{aligned}$$

Alternatively, we can use the **distributive law** and remove the brackets by multiplying each term inside the brackets by the term outside the brackets.

$$\begin{aligned} 3 \times (8 + 5) &= 3 \times 8 + 3 \times 5 \\ &= 24 + 15 \\ &= 39 \end{aligned}$$

A mobile phone company charged 17 toea per 20-second block. This comprised 11 toea production cost and 6 toea profit. The accountant described this mathematically as $1 \times (11 + 6) = 17$.

4 Write a mathematical expression the accountant could use to describe the following calls and use the distributive law to calculate the cost of each. Show how you worked out your answers.

- **a** Three blocks of time charged at a rate of 19 toea production cost and 10 toea profit per block.
- **b** Two blocks of time charged at a rate of 24 toea production cost and 3 toea profit per block.
- **c** Six blocks of time charged at a rate of 10 toea production cost and 4 toea profit per block.
- **d** Five blocks of time charged at a rate of 11 toea production cost and 5 toea profit per block.
- **e** Four blocks of time charged at a rate of 9 toea production cost and 7 toea profit per block.

Help Box

The **distributive law** is particularly useful in algebra where the terms inside the brackets are not always like terms. We use the distributive law to remove the brackets by multiplying each term inside the brackets by the term outside the brackets. This process is called **expansion**.

For example:

$$\begin{aligned} 4(m + n) &= 4 \times (m + n) \\ &= 4 \times m + 4 \times n \\ &= 4m + 4n \end{aligned}$$

5 Expand each of the following.

a	$5(x+y)$	b	$2(s+t)$	c	$7(a+b)$	d	$4(g+h)$
e	$2(p+5)$	f	$4(c+7)$	g	$9(a+1)$	h	$3(6+m)$
i	$6(2+q)$	j	$8(3+r)$	k	$3(7+e)$	l	$(g+3)5$
m	$(t+6)2$	n	$(j+4)2$	o	$6(5+d)$	p	$(g+2)9$

The distributive law also applies to subtraction.

6 Expand each of the following.

a	$4(p-y)$	b	$5(m-g)$	c	$3(q-p)$	d	$4(t-p)$
e	$8(w-4)$	f	$2(f-7)$	g	$5(x-5)$	h	$5(5-y)$
i	$4(4-j)$	j	$6(4-e)$	k	$8(4-g)$	l	$(k-5)9$
m	$(h-5)7$	n	$(n-3)4$	o	$2(1-r)$	p	$(t-7)3$

7 Expand each of the following.

a	$m(p+h)$	b	$j(m+n)$	c	$k(r+p)$	d	$h(l+p)$
e	$x(y-5)$	f	$k(t-4)$	g	$a(y-4)$	h	$x(8-y)$

8 Expand each of the following.

a	$4(2p+h)$	b	$3(2x+r)$	c	$3(g+3e)$	d	$(y+4b)5$
e	$2(3t-4)$	f	$2(5k-7)$	g	$4(5-3d)$	h	$(g-2m)9$

The distributive law applies to pronumerals as well as numerals.

9 Expand each of the following. The first one has been done as an example.

a	$m(3+2b)=3m+2bm$	b	$x(2-3y)$
c	$(2a+5)b$	d	$(3k-1)n$
e	$(4-5y)\,x$	f	$(4k-3)m$
g	$s(6p-1)$	h	$g(5+2h)$

Expand $2x(4x-1)$.

Lesson 7 Keep it simple

Algebra provides a shortcut method of writing a set of instructions for doing calculations.

1 Write an algebraic expression for each of the following.

a The sum of a and b is multiplied by 5.

b n is added to 4 and the result is multiplied by 3.

c The difference between 12 and x is multiplied by 4.

d 6 is added to y and the result is multiplied by 7.

e The sum of 3 and y is divided by 2.

f k is divided by m and 4 is added to the result.

2 Write three other descriptions similar to those in question 1, then write the algebraic expression for each.

3 On Monday Jenny made m standard 30-second block calls and n information calls on her mobile phone. She was charged 20 toea for each call including flagfall. Write an algebraic expression to describe:

- **a** the total number of calls Jenny made on her mobile on Monday
- **b** the amount she was charged in total for those calls

4 A mobile phone company charged a toea per 30-second phone block and b toea per information call. A customer made 21 standard 30-second block calls and 5 information calls in the month. Write an algebraic expression to describe:

- **a** how much the customer was charged for information calls that month
- **b** what the customer's bill was that month for standard 30-second block calls

Challenge

Write two stories similar to those in questions 3–5 that could each be represented by the expression $5(a + b)$.

5 A group of friends hire a mobile phone to take with them on a holiday. There are x females and y males in the group. Each person pays 3 kina for their share of the phone costs. Write an algebraic expression to describe the total amount of money collected from the group.

6 If m is the number of TXT messages that Alice sent in a week and n is the number that Edna sent, the equation $m - 4 = n$ means that Edna sent 4 less TXT messages than Alice that week. Explain the meaning of each of the following equations.

a	$n = m$	**b**	$n = 2m$	**c**	$m = n + 5$	**d**	$m = 2n + 3$
e	$m - 6 = n$	**f**	$n = m - 2$	**g**	$m = n - 3$	**h**	$m = 2n$

Earlier we found that the distributive law allows us to expand algebraic expressions, and that like terms can be grouped together. So it is possible to expand algebraic expressions and then simplify them by grouping the like terms.

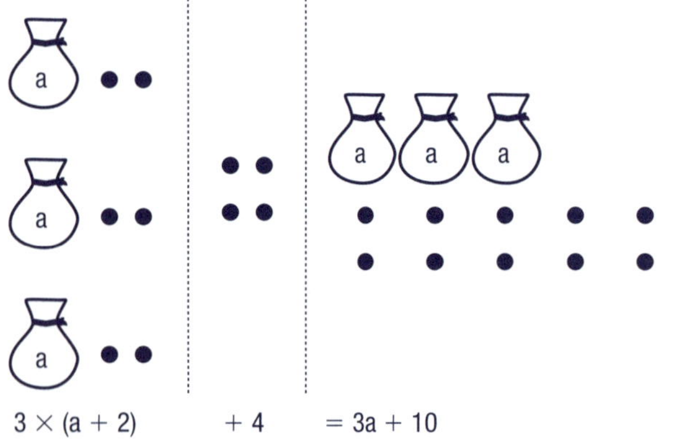

Help Box

The expression $3(a + 2) + 4$ can be expanded using the distributive law:

$$3(a + 2) + 4 = 3a + 6 + 4$$

and then simplified to $= 3a + 10$

The expression $5(2x + y) + 3x$ can be expanded using the distributive law:

$$5(2x + y) + 3x = 10x + 5y + 3x$$

and then simplified to $= 13x + 5y$

7 Expand and simplify.

a	$3(a + 5) + 4$	**b**	$4(x - 3) + 2$	**c**	$2(k + 4) - 6$	**d**	$7(m - 4) - 8$
e	$5(2b + 1) + 6$	**f**	$3(2p - 7) - 11$	**g**	$2 + 4(b + 6)$	**h**	$5 - 2(4 - h)$
i	$4 + 3(x - 1)$	**j**	$3 + 4(2a + 7)$	**k**	$8(2x + 2) - 4$	**l**	$5 + 2(a + 3 - b)$

8 Simplify the following expressions by expanding using the distributive law then collecting like terms.

a $3(4 + l)$
b $3(7 + 5p) - 4p$
c $3k(3 - 2m) + 4k$
d $4z(3 - 2h) + 4z$
e $9(3s - 4) + 2(s + 3)$
f $10 + 3(r + 2) - r + 3$
g $5 + h(6 - c) + h - 2c$
h $4(m - 3c + 3) + 2m$
i $4k + 7(4 - k + 5)$

Lesson 8 Substitution

You have learned that an algebraic expression can have different values depending on the value of its pronumerals. If we know the value of a pronumeral, we can substitute it with a number and solve the equation. This process is called **evaluating** the algebraic expression.

Help Box

To **evaluate** an algebraic expression, substitute the pronumeral with the given number.

For example, given that $x = 2$ and $y = 3$ we can substitute the expression $x + y$ with $2 + 3$ and find the solution is 5.

Remember to put in the multiplication signs when evaluating.

For example, given that $m = 4$ and $p = 5$, we can substitute the expression $2m + 3p$ with $2 \times 4 + 3 \times 5$ and find the solution is 23.

1 a Evaluate the algebraic expression $m + 5$ if m is substituted by 3.

b Evaluate the algebraic expression $k - 4$ if k is substituted by 8.

c Substitute x with 2 to evaluate the expression $4x + 3$.

d If $h = 8$, evaluate the expression $37 - 4h$.

e Given that $t = 3$, find the value of $3t + 5$.

f Substitute g with 6 and p with 2 to find the value of $3g + 2p$.

2 Substitute a with 6 and b with 4 to evaluate the following expressions.

a $4a$
b $5b$
c $a - b$
d $a + b$
e $\frac{a}{b}$
f $\frac{12}{b}$
g $\frac{a}{18}$
h $2a + 4b$
i $\frac{2a + b}{4}$
j $3a + \frac{8}{b}$

3 Substitute $a = 3$ and $b = 5$ into each of the following expressions to solve the equations.

a $3a + 2b =$
b $5a + 3b =$
c $6a - 2b =$
d $3a + b =$
e $ab =$
f $2ab =$
g $\frac{5a}{b} =$
h $\frac{20a}{2b} =$

4 Substitute $x = 8$ and $y = 3$ to solve each of the following expressions.

a $2x + 3y$
b $11 - 2y$
c $2x + 4$
d $13 - 3y$
e $2(x + 4)$
f $y(3 + 4)$
g $4(x + y)$
h $5x - \frac{x}{2}$
i $\frac{24}{4y} + x$
j $\frac{30}{2y} - 5 + x$

5 Use the values in the table to evaluate the following expressions.

v	w	x	y	z
2	3	4	5	6

a $v + w$
b vx
c $w + 7y$
d $4z - w$
e $3(v - 4)$
f $x(2v + y)$
g $4xyz$
h wxz
i $wx + 5$
j $9v - 3x$
k $2z + 3w - x$
l $4v(\frac{6w}{2})$

6 Copy the following tables into your workbook and use the equation to find the missing numbers.

a $x + 7 = y$

x	y
3	10
11	
19	
	38
	51

b $4c + 3 = h$

c	h
10	43
7	
2	
1.5	
0.2	

c $\frac{d}{2} = t$

d	t
8	4
14	
26	
3	
0.16	

7 Write the equation represented by the following table of values.

p	q
3	10
7	22
10	31
15	46
100	301

Lesson 9 Factorisation

Earlier we saw how a mobile phone bill can be broken down to show the charge per 30-second block.

A number that divides exactly into another number with no remainder is called a **factor**. In this phone charge example, 3 is a factor of 75 because 3 'goes into' 75 (25 times) with no remainder. The other factors of 75 are 1, 5, 15, 25, 75.

The smallest factor of a number is always 1 and the largest factor is always the number itself.

1 List the factors of the following numbers.

a 9 b 15 c 20 d 12 e 36

f 52 g 80 h 69 i 240 j 97

Help Box

The **highest common factor** (HCF) is the highest number that appears in both the lists of factors for two different numbers. To find the HCF first list all the factors of each number, then find the largest number that appears in both lists.

For example, to find the HCF of 54 and 36:

List the factors of 54 $1 \times 54, 2 \times 27, 3 \times 18, 6 \times 9$, so the factors of 54 are (1, 2, 3, 6, 9, 18, 27, 54)

List the factors of 36 $1 \times 36, 2 \times 18, 3 \times 12, 4 \times 9, 6 \times 6$, so the factors of 36 are (1, 2, 3, 4, 6, 9, 12, 18, 36)

So the HCF of 54 and 36 is 18.

2 a Write the factors of 18.

b Write the factors of 42.

c List the common factors of 18 and 42.

d What is the highest common factor of 18 and 42?

3 Find the highest common factor of:

a 5 and 25 b 16 and 28

c 32 and 48 d 40 and 15

e 42 and 70 f 9 and 39

g 8, 50 and 70 h 32, 56 and 80

i 6, 12 and 30 j 33, 55 and 132

k 75, 30 and 123 l 28, 91 and 63

4 Write two numbers that have a HCF of:

a 7 b 3 c 12 d 1

5 Factorise the following expressions.

a $2g + 12$ b $4t + 16$ c $3d + 15$

d $8w + 4$ e $6s - 3$ f $24k - 12$

g $15 + 35x$ h $18y - 45$ i $110 - 22b$

j $3h + 39$ k $42t - 12$ l $80 + 25p$

6 Factorise the following expressions.

a $5a + 5b$ b $21h + 39d$

c $24x - 3y$ d $28w - 14g$

e $15t + 25g$ f $18mp - 20mp$

g $24kn + 36kn$ h $7mn + 14m$

i $6bt + 2b$ j $21gw - 14g$

k $6wv + 2w$ l $90vr + 5r$

Help Box

Algebraic expressions can also be broken down into **factors**. Think of factorising as the opposite of expanding.

For example, to break the expression $9m + 15$ into factors:

Step 1 Break the terms into their HCF $9m + 15 = 3 \times 3 \times m + 3 \times 5$

Step 2 Put the HCF outside the brackets $= 3 \times (3 \times m + 5)$

Step 3 Remove the multiplication symbol $= 3(3m + 5)$

Step 4 Expand the answer to check its accuracy $= 9m + 15$

7 Use factors to help solve the following problems. The first one has been done for you.

a $38 \times 18 = 38 \times 9 \times 2$

$$\begin{array}{r} 38 \\ \times 9 \\ \hline 342 \\ \times 2 \\ \hline 684 \\ \hline \end{array}$$

b 23×16

c 53×32

d 68×45

e 242×18

Challenge

Use factors to calculate:

a $645 \div 15$ b $252 \div 18$ c $1024 \div 16$

Lesson 10 Mobile phone debt

In Grade 7 we located negative numbers on the number line.

Some people link their mobile phone plan to their bank account. The bank pays the phone bill from the person's account, or lends them the money if they don't have enough in their account when the phone bill arrives.

1 Use the number line at the right of the page to help you calculate the final account balance of the following:

- **a** a savings account balance of K12 followed by a withdrawal to pay a phone bill of K10
- **b** a savings account balance of K6 followed by a withdrawal to pay a phone bill of K9
- **c** a savings account balance of K7 followed by a withdrawal to pay a phone bill of K15
- **d** a savings account balance of K10 followed by a withdrawal to pay a phone bill of K14

2 Calculate the account balance after the following transactions (assume the balance was zero to begin):

- **a** a deposit of K10 followed by a withdrawal of K15
- **b** a deposit of K15 followed by a withdrawal of K9 and then another withdrawal of K11
- **c** a deposit of K8 followed by a withdrawal of K12
- **d** a deposit of K7 followed by a withdrawal of K12 and then a deposit of K5
- **e** a deposit of K9 followed by a withdrawal of K6 and then a deposit of K4
- **f** a deposit of K3 followed by a withdrawal of K14 and then a deposit of K6

3 Write a number sentence to describe and solve each situation.

- **a** I have K8 in my savings account then I deposit K5 before withdrawing K7 to pay my phone bill.
- **b** My savings account is empty and I owe the bank K15. I pay back K11 to the bank and soon after they lend me another K8 to pay my next phone bill.
- **c** I have K25 in my savings account and I withdraw K12 to pay my phone bill. I deposit a further K5 in the account and then withdraw K22 to buy a new phone.

4 Use a number line to solve:

- **a** $-4 + +5$
- **b** $+6 + -2$
- **c** $-8 + -5$
- **d** $-7 + -4$
- **e** $8 + -3$
- **f** $-9 + +4$
- **g** $+3 + -4$
- **h** $-13 + +8$
- **i** $+8 + -16$
- **j** $+5 + -2 + -7$
- **k** $-4 - +5 + -3$
- **l** $-5 + +4 - +7$
- **m** $-24 + -6 + +9$
- **n** $-12 + -8 + +13$
- **o** $+24 + -30 + -7$
- **p** $-3 + -32 + -18$

Subtracting negative numbers

5 Copy the addition equations below into your book.

- **a** What is the pattern in the second number of each equation?
- **b** What is the pattern in the answers to each equation?
- **c** Continue the pattern by writing the next three equations.

6 Copy the subtraction equations below into your book.

- **a** What is the pattern in the second number of each equation?
- **b** What is the pattern in the answers to each equation?
- **c** Continue the pattern by writing the next equation at both ends of the table.

$3 + (+3) = +6$
$3 + (+2) = +5$
$3 + (+1) = +4$
$3 + (0) = +3$
$3 + (-1) = +2$
$3 + (-2) = +1$
$3 + (-3) = 0$
$3 + (-4) = -1$
3 + ……..

3 – ……..
$3 - (+4) = -1$
$3 - (+3) = 0$
$3 - (+2) = +1$
$3 - (+1) = +2$
$3 - (0) = +3$
$3 - (-1) = +4$
$3 - (-2) = +5$
3 – ……..

11 10 9 8 7 6 5 4 3 2 1 0 –1 –2 –3 –4 –5 –6 –7 –8 –9

7 Rewrite the following subtractions as their equivalent addition equation and solve.

a 6 – +5　　b 13 – –6
c –8 – +4　　d –42 – –6
e –12 – +7

8 Calculate:

a +5 – +3　　b –7 – +4　　c +2 – –3
d –9 – –6　　e 2 – +4　　f 9 – –2
g –4 – +3　　h –6 – +4　　i –8 – –6
j +4 – +1

Help Box

When dealing with subtraction of directed numbers we **add the opposite** of the **number we want to subtract**.

For example 9 – +3	becomes	9 + –3	Answer = 6
9 – –3	becomes	9 + +3	Answer = 12
–9 – +3	becomes	–9 + –3	Answer = –12
–9 – –3	becomes	–9 + +3	Answer = –6

There are some rules that help to simplify addition and subtraction of directed numbers. Already you know that it is not always necessary to use the + sign when writing positive numbers. It is accepted that 6 has the same value as +6.

This means that +5 + +6 has the same value as 5 + 6.

So + + is the same as +.

The calculations you have done have also shown that +5 + –6 has the same value as 5 – 6.

So + – is the same as –.

You now know that because subtraction is the opposite of addition, +5 – +6 has the same value as +5 + –6 which has the same value as 5 – 6.

So – + is the same as –.

Because subtraction is the opposite of addition, you also now know that +5 – –6 has the same value as +5 + +6 which has the same value as 5 + 6.

So – – is the same as +.

9 Copy this summary into your workbook.

Summary of rules

(+ +) = +, (– –) = + : Two like symbols make a positive

(+ –) = –, (– +) = – : Two unlike symbols make a negative

10 Simplify the following before calculating the answer.

a 13 + +15　　b 2 – +18
c 52 + –9　　d 23 – –16
e –12 + +5　　f –32 – +5
g –13 + –23　　h –9 – –21
i –12 + –13　　j –11 – –9

11 Copy into your book and write True or False for each example.

a 13 + –5 = 13 – +5
b 24 + +6 = 24 + 6
c 18 – +3 = 18 + 3
d –11 + –5 = 11 – 5
e – 8 – –8 = + 8 + 8
f 7 – +4 = 7 – 4
g 5 – –8 = 5 + 8
h –7 + +6 = –7 + 6
i +9 – +10 = 9 – 10
j –15 – +9 = –15 + 9

Learning Unit 2 Wheels

Strand: Space and Shape

Area	Outcome 8.2.5	Investigate the area of circles
Shapes	Outcome 8.2.9	Make physical models of circles and investigate their properties

Strand: Chance and Data

Sets	Outcome 8.4.2	Use sets to solve problems from real life
Estimation	Outcome 8.4.6	Identify and select appropriate estimation strategies

Lesson 1: Introduction	Circles in our daily life
Lesson 2: Features of circles	Identifying circumference, diameter, radius, segment and arc
Lesson 3: One turn of the wheel	Relating circumference, radius, diameter and π Introduction of the formulae $C = \pi D$ $C = 2\pi r$
Lesson 4: Applying the formulae for circumference	Solving problems involving diameter, radius and circumference
Lesson 5: Whipper snippers and estimating area of circles	Drawing simple scale drawings Estimating area of circles Counting grid squares to find area of circles
Lesson 6: Investigating area of circles	Estimating area Investigating inscribed squares
Lesson 7: Finding the formula	Using sectors to find a formula Applying πr^2
Lesson 8: Calculating the area of a circle	Using the formula for area
Lesson 9: Part circles	Compound area in real-life situations
Lesson 10: Sets	Using Venn diagrams

Lesson 1 Introduction

This unit examines some of the features of circles by exploring the wheels in our everyday lives. You will be able to identify some of the measurement relationships common to all circles. You will also investigate ratios and revise the use of sets to classify items and solve problems.

The wheel is possibly the most important invention in human history. It significantly improved our ability to transport goods and provided us with the means to travel greater distances in a shorter time. From the humble wheelbarrow with its single wheel to the enormous road trains with multiple wheels, we are arguably now more dependent on this one invention than any other in history.

1 Where do you see wheels in your daily life?

2 How many students in the class own a bicycle?

3 Why do wheels on most transport vehicles have tyres?

4 In what ways would life be different for this market gardener if he did not have a wheelbarrow?

5 **a** On a piece of card, trace around a circular object.

b Cut the circle out and estimate the distance around its outside. Write your estimation on the circle.

c Wind a piece of string around the circle, and then hold the string against a ruler to find the distance. Write this measurement on the other side of your circle.

6 Use the circle you made above as a template to create this pattern.

Lesson 2 Features of circles

The distance around a circle is called the **circumference**. Measuring the circumference of a circle can be difficult compared to finding the perimeter of a straight-sided shape. In your previous lesson you measured the circumference of a circle using a piece of string. Now try another method.

Step 1: Make a mark on the edge of a circular object such as a tin or small bottle.

Step 2: Place the mark on the zero of your ruler and roll the object along the ruler one full turn. Note the point on the ruler where the turn ended. The distance from zero to this final point is the circumference of the circle.

Step 3: Count how many times the circular object will fit edge to edge along that same distance on the ruler.

1 Copy this sentence into your book and complete it based on what you found.

The circumference of the object was ______ cm. The object fitted along a line equal in length to its circumference about ___ times.

The distance across the centre of any circle is called its **diameter**. The **radius** is the distance from the centre of the circle to the circle's edge. It is half the length of the diameter.

radius
diameter
circumference

2 a What is the diameter of the circular object you used in question 1?

b What is its radius?

3 Explain how to calculate the diameter of a circle if you know the radius.

4 The circles in the diagram are **concentric** because they all have the same centre point. Measure the radius of each circle and calculate the diameter. Then use a piece of string to measure the circumference of each. Record your results in a table as shown.

Circle	A	B	C	D	E
Radius (cm)					
Diameter (cm)					
Circumference (cm)					

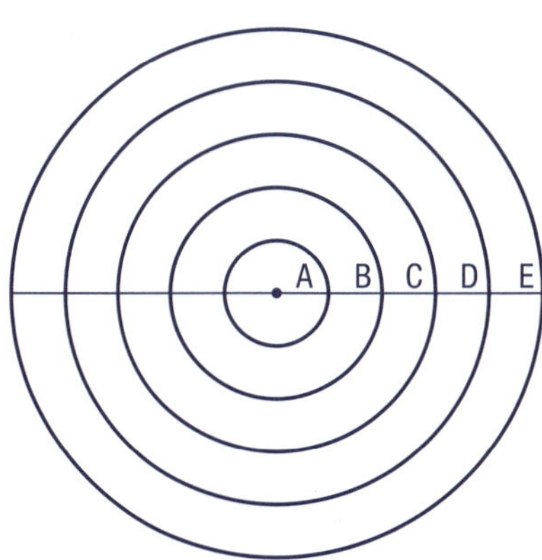

5 a What relationship did you notice between the diameter and the circumference for each of the circles in question 4?

b What relationship did you notice between the radius and the circumference for each of the circles in question 4?

6 Calculate the diameter of a circle with a radius of:

a 44 cm b 275 mm
c $\frac{3}{4}$ m d 2.8 m

7 Calculate the radius of a circle with a diameter of:

a 400 mm b 38 cm
c $2\frac{1}{2}$ m d 4.6 m

8 Estimate the circumference of a circle with a diameter of:

a 20 cm b 35 cm
c 250 mm d 85 cm

9 Estimate the circumference of a circle with a radius of:

a 100 mm b 15 cm
c 25 cm d 2.4 m

10 a Use a pair of compasses to draw a circle with a radius of 14 cm.

b Cut the circle out and mark and label the centre, radius, diameter and circumference.

c Write the length of the radius, diameter and circumference on the appropriate part of the circle.

An **arc** is a part or portion of a circle. This diagram shows the arc between R and S.

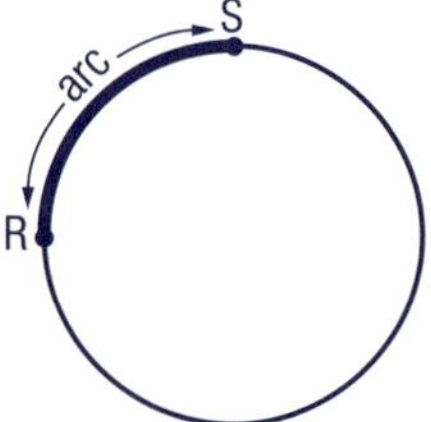

11 Use a piece of string and a ruler to help you mark an arc of 20 cm on the circle you cut out. Label the arc on your circle appropriately.

A **circle segment** is an area of a circle that is 'cut off' from the rest of the circle. It excludes the circle's centre.

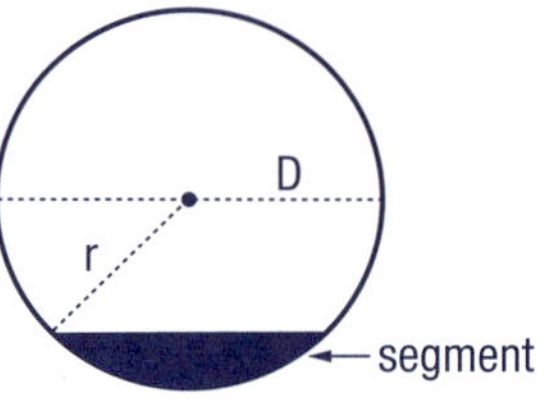

12 Colour and label a segment on your circle.

Lesson 3 One turn of the wheel

Use your knowledge of the relationships between radius, diameter and circumference of circles to answer these questions about wheels.

1 A wheel has a diameter of 95 cm.

- a What is its radius?
- b Estimate its circumference.

2 A wheel has a circumference of about 63 cm.

- a Estimate its diameter.
- b What might its radius be?

3 The radius of a wheel is 30 cm.

- a What is its diameter?
- b What might its circumference be?

Challenge

What happens to the circumference of a circle if we double its diameter?

What happens to the diameter of a circle if its circumference is enlarged five times?

The circumference of any circle is about three times its diameter. The special symbol π stands for the number of times the diameter of a circle fits into its circumference. The exact figure cannot be found because it has an infinite number of decimal places. Generally the numbers 3.14 or $3\frac{1}{7}$ are accepted as close approximations of π.

Help Box

To calculate the circumference of a circle when the diameter is known use the formula $C = \pi D$.

For example, calculate the circumference of a circle with a diameter of 10 cm.

$$C = \pi D$$
$$C = 3.14 \times 10$$
$$= 31.4 \text{ cm}$$

The circumference of the circle is approximately 31.4 centimetres.

4 Use the formula to find how far a wheel will travel in one revolution if it has a diameter of:

a 1 m b 20 cm c 50 cm d 2.5 m

5 How far will a wheel with a diameter of 180 cm travel in 100 revolutions?

6 The wheel on a tractor makes 50 revolutions moving from one end of the field to the other. If the wheel has a diameter of 2 m, how long is the field?

7 The world's biggest truck is the Caterpillar 797B used in the mining industry. Each tyre has a diameter of 4 m. How far does the truck move in 10 wheel revolutions?

8 In London, a giant Ferris wheel known as the London Eye has a diameter of 135 metres. How far do people travel in one revolution on the London Eye?

Help Box

To calculate the circumference of a circle when the radius is known, use the formula $C = 2\pi r$ (the radius doubled is equivalent to the diameter).

For example, to calculate the circumference of a circle with a radius of 5 cm:

$$C = 2\pi r$$
$$C = 2 \times 3.14 \times 5$$
$$= 31.4 \text{ cm}$$

The circumference of the circle is approximately 31.4 centimetres.

9 Use the formula to work out the circumference of a tyre with a radius of:

a 3 m b 35 cm

c 1.5 m d 75 cm

10 How far will a wheel with a radius of 50 cm travel in 300 revolutions?

Challenge

The city of Shanghai in China hopes to build a Ferris wheel with a 170 m diameter.

What difference in length will there be between:

a the radius of the London Eye and the Shanghai Wheel?

b the circumference of the London Eye and the Shanghai Wheel?

Lesson 4 Applying the formulae for circumference

When working with wheels there are many times when knowing about π makes it easier to calculate solutions.

1 A bicycle wheel has a diameter of 80 cm. Which tyre of these circumferences will be the best fit? Explain your answer using π.

A 1.5 m B 2.5 m C 3.5 m

2 A wheel rim with a circumference of 2 metres needs some spokes. (Spokes are the thin metal rods that provide support from the centre of the wheel to the rim.) Which spokes would be closest to the length required? Explain your answer using π.

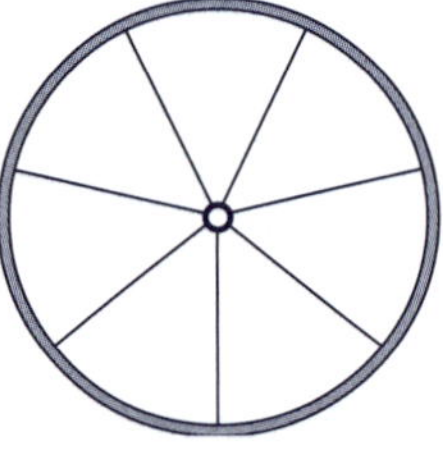

A 33 cm B 43 cm C 53 cm

3 A bike spoke measuring 40 cm would be most suitable for use on a bike wheel with a diameter of:

A 1.2 m B 2.4 m C 80 cm?

Explain your answer using π.

4 If we rolled some different sized wheels along a flat surface, which one would roll furthest in each rotation? Explain your answer using π.

A 

B 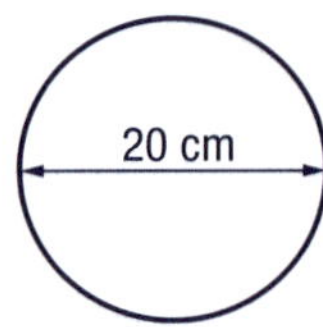

C 

5 Estimate the circumference of a wheel with a diameter of:

a 90 mm b 28 cm

c 3 m d 1.2 m

6 Use the formula to calculate more accurate answers to question 5.

7 Estimate the diameter of a wheel with a circumference of:

a 21 cm b 4.2 m

c 75 cm d 1.5 m

8 Use the formula $D = \frac{C}{\pi}$ to calculate more accurate answers to question 7.

9 Calculate the radius of a wheel with a diameter of:

a 8 cm b 2 m c 2.4 m d 38 mm

10 Calculate the diameter of a wheel with a radius of:

a 3.24 cm b 8 m c 150 mm d 21 m

11 Use a formula to calculate the circumference of the following circles.

a
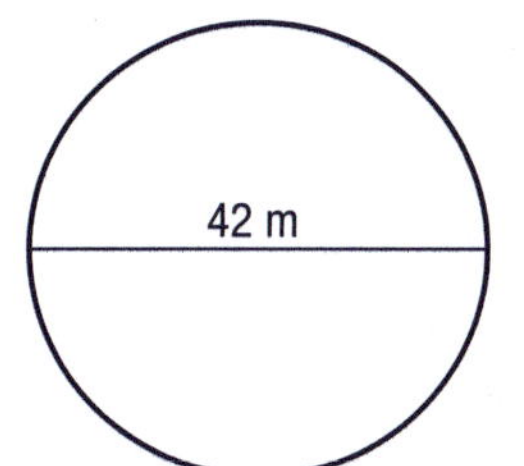

b
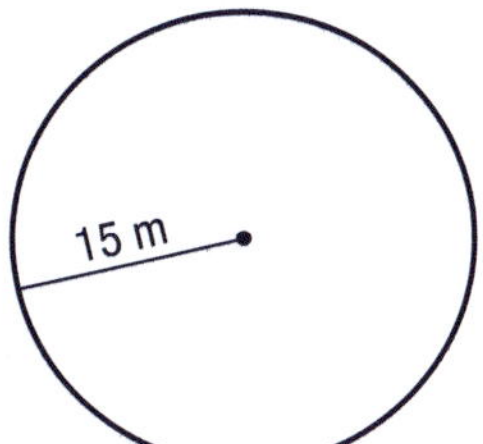

c
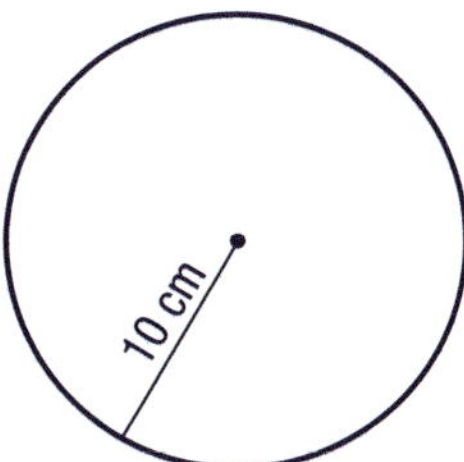

d
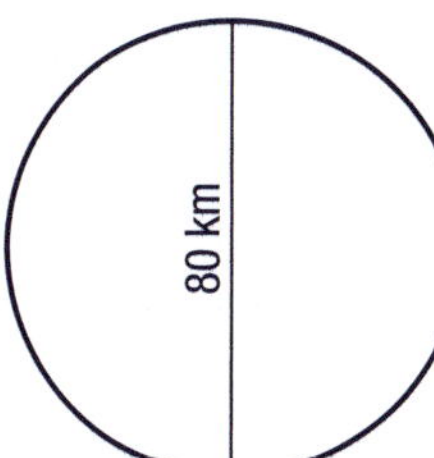

e
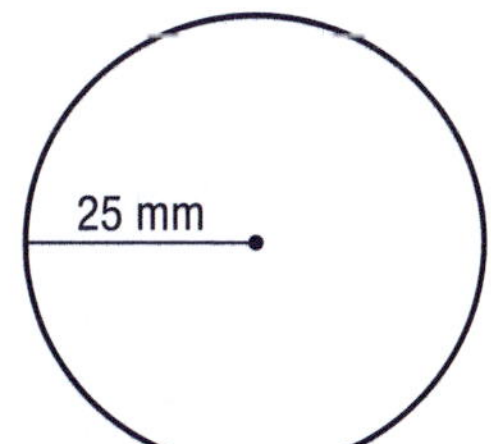

f
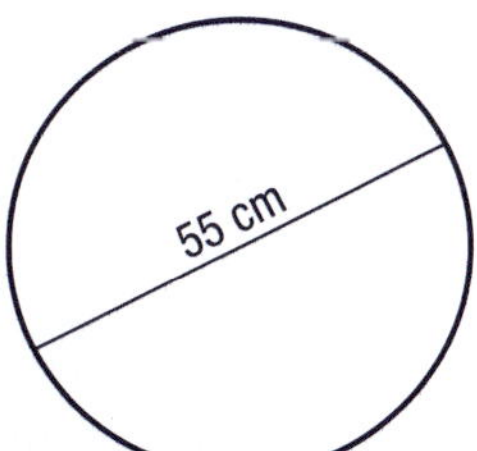

g
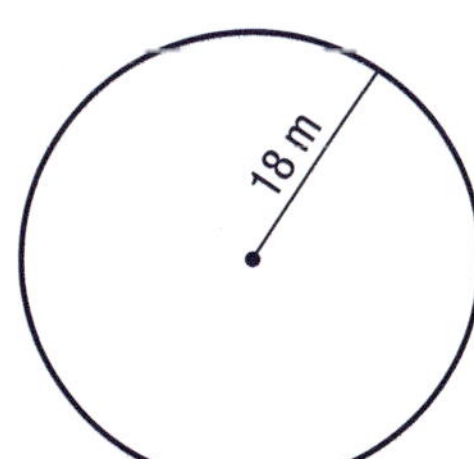

h
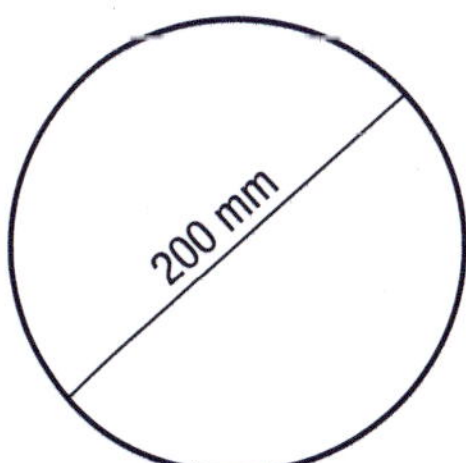

12 Calculate the circumference of a wheel with:

a a diameter of 20 cm b a radius of 15 mm c a diameter of 10 cm

d a radius of 18 cm e a diameter of 20 cm f a radius of 4.2 km

Lesson 5 Whipper snippers and estimating area of circles

In some countries people use machines known as 'whipper snippers' to cut grass, as shown. Basically, a whipper snipper has an engine that spins a nylon thread at high revolutions. The nylon thread hitting the blades of grass at high speed severs the grass at the point of impact.

These are the lengths of nylon thread attached to three different sized whipper snippers.

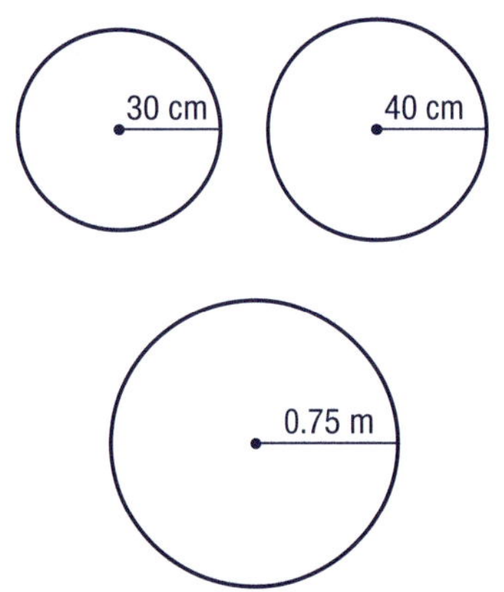

1 Draw a scale diagram on grid paper to show the area of the cutting circle for each of the whipper snippers. Each centimetre on your grid paper should represent 10 cm in real life.

2 Estimate the area enclosed within the circle for each of the different sized whipper snippers.

3 Find the approximate area of each cutting circle by counting squares on the grid paper. You will need to decide how to deal with parts of squares that are inside a circle.

4 Draw rectangular shapes that enclose an area equal to each of the cutting circles. Label your diagram to show the dimensions of each rectangle.

5 Discuss with a friend why estimating the area of circles seems harder than estimating the area of rectangles.

Challenge

Does there appear to be a pattern between the length of the plastic cord and the area of the cutting circle? If so, describe what the pattern appears to be.

Lesson 6 Investigating area of circles

In the last lesson you drew circles on grid paper and counted squares to estimate the area of circles. We can estimate the area of circles using some other methods. One other way of estimating the area of a circle is to use the area of other shapes.

Inscribed means fitting a shape exactly inside a circle with its vertices just touching the edge of the circle. Calculating the area of inscribed shapes allows us to approximate the area of the circle. First we can use our knowledge of triangles to find the area of a triangle inscribed in a circle.

Help Box

The circle on the right has a diameter of 5 cm.

In the diagram below, an angle of 45° is measured and drawn at each end of the diameter. The angles meet to form a right angle at the circumference. Note that the dotted line from the centre of the circle to the vertex of the triangle forms a radius with a length of 2.5 cm.

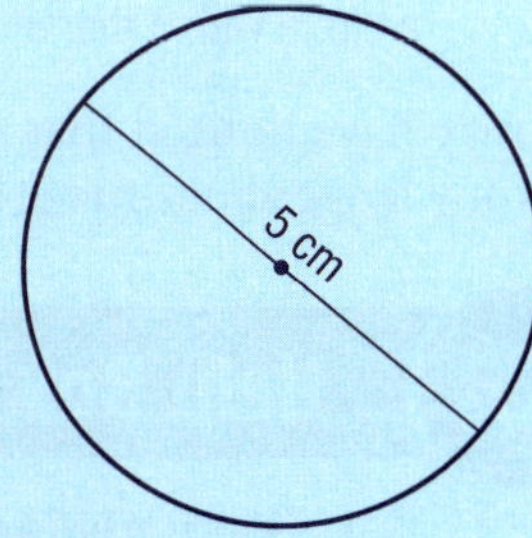

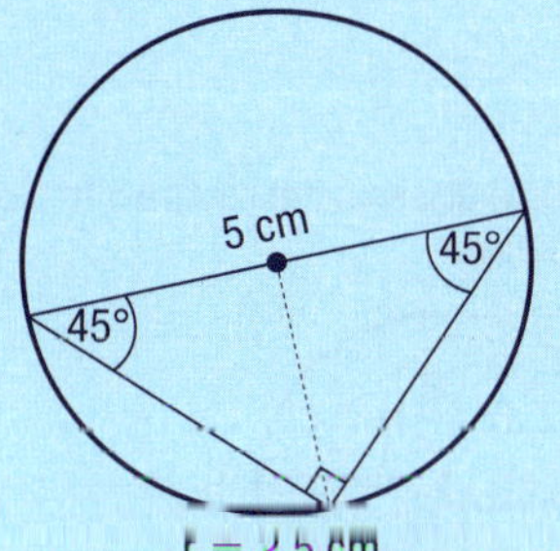

The formula for finding the area of a triangle is Area (A) $= \frac{\text{Base} \times \text{Height}}{2}$

The area of the right-angled triangle is therefore

$$A = (5 \times 2.5) \div 2$$
$$= 12.5 \div 2$$
$$= 6.25\ \text{cm}^2$$

An identical right-angled triangle can be drawn on the other side of the diameter.

The total area of the two triangles is 12.5 cm^2.

We can therefore estimate that the area of the circle is a little more than 12.5 cm^2.

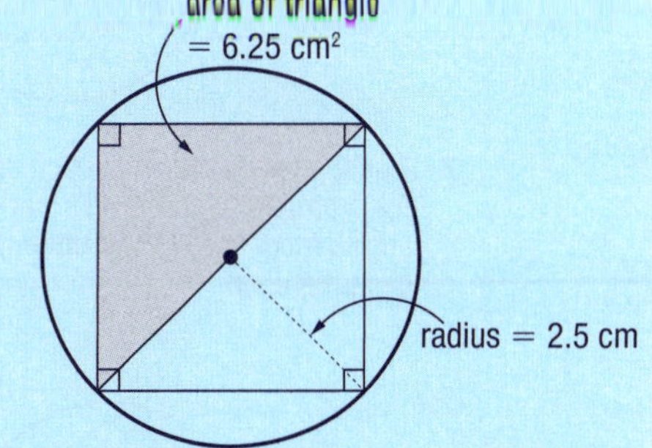

1 a Follow these steps to show how to estimate the area of a circle using an inscribed triangle.

- Use a compass to draw a circle with a radius of 4 cm.
- Rule a line across the diameter.
- Measure a 45° angle at each end of the diameter.

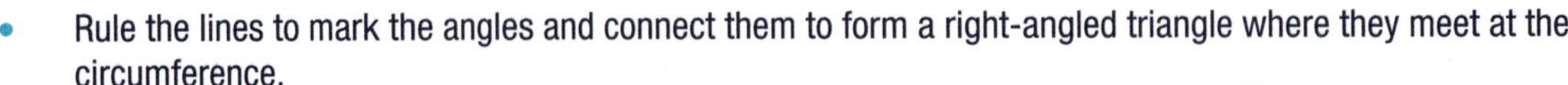

- Rule the lines to mark the angles and connect them to form a right-angled triangle where they meet at the circumference.
- Mark in the radius using a dotted line and label its length.
- Calculate the area of the inscribed triangle.
- Measure and draw the identical triangle on the other side of the diameter.
- Estimate the area of the circle by doubling the area of the triangle.

b Copy and complete the sentence: The estimated area of a circle with a radius of 4 cm, using the inscribed triangle method, is a little more than ______ cm^2.

2 Draw a triangle inscribed inside a circle to estimate the area of a circle with a diameter of 12 cm.

3 Use the inscribed triangle method to estimate the area of the following circles. Show your working out. You may not need to actually draw the circle at this stage.

a a circle with a radius of 5 cm

b a circle with a radius of 20 cm

c a circle with a diameter of 6 m

d a circle with a radius of 3 m

You may have noticed that in drawing two identical inscribed triangles in a circle, a square was formed. Using our knowledge about the area of squares is another way of estimating the area of a circle.

Help Box

1 This circle has a diameter of 10 cm.

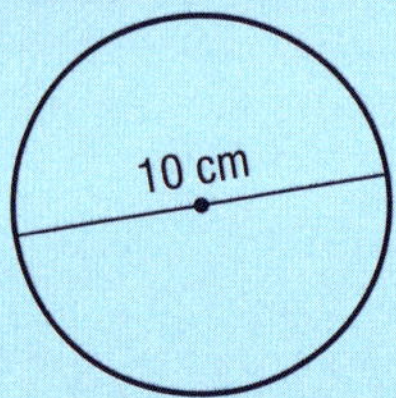

2 A right angle is drawn at each end of the diameter. Align the protractor with the 45° mark on the diameter. The angles meet to form two other right angles at the circumference.

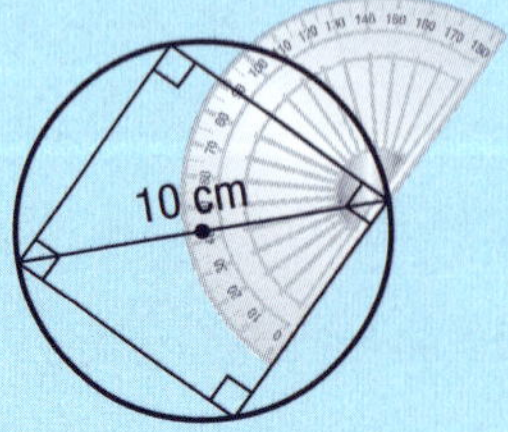

3 Adding another diagonal dissects the square into four small right-angled triangles.

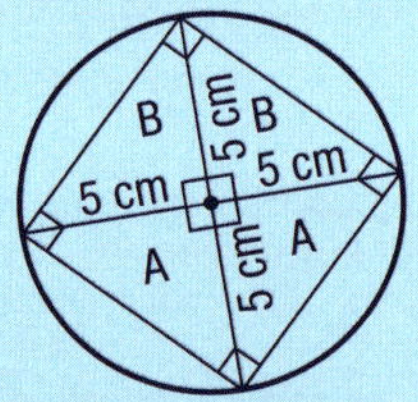

4 The diagram shows the triangles can be paired to make two small squares. The side length of each square is equal to the radius of the circle (5 cm).

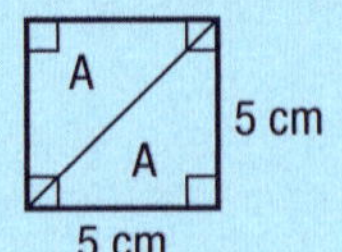

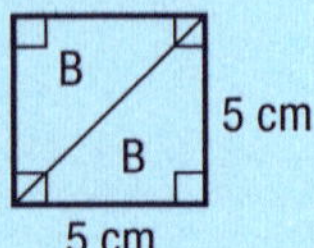

The total area of the inscribed square is $2 \times (r \times r)$ or $A = 2r^2$.

We can therefore estimate the area of the circle is a little more than $(2 \times 5^2) = 50$ cm^2.

4 **a** Follow these steps to show how to estimate the area of a circle using an inscribed square.

- Use a compass to draw two circles each with a radius of 8 cm on a piece of loose paper.
- Cut out both circles and paste one into your workbook.
- Fold the other circle in half and then in half again.
- Open out and write the length of the radius along each fold line.
- Rule a line between each fold line to form an inscribed square that intercepts at the circumference.
- Cut out the four small triangles contained in the circle and regroup them to make two small squares with a side length of 8 cm.
- Paste the two squares and the other remaining curved pieces into your book.

b Copy and complete the sentence: The estimated area of a circle with a radius of 8 cm, using the inscribed square method, is a little more than _____ cm^2.

5 Use the square method to estimate the area of the following circles. Show your working out. You may not need to actually draw the circle at this stage.

a a circle with a radius of 7 cm

b a circle with a radius of 11 cm

c a circle with a diameter of 16 m

d a circle with a radius of 100 m

Finding the formula

Another way of estimating the area of a circle is to cut the circle into segments and rearrange the segments into a rough rectangle.

The circumference of any circle is equal to $2\pi r$.

The length of the curved edge of the shaded region in this circle is one half of the circumference.

Half the circumference can be calculated as $2\pi r \div 2$ (which equals πr).

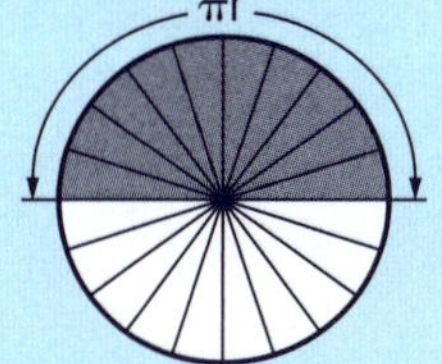

The circle can be folded into sectors, then the sectors cut out and rearranged to form a shape very similar to a rectangle.

The rectangle's length is πr and its height is r.

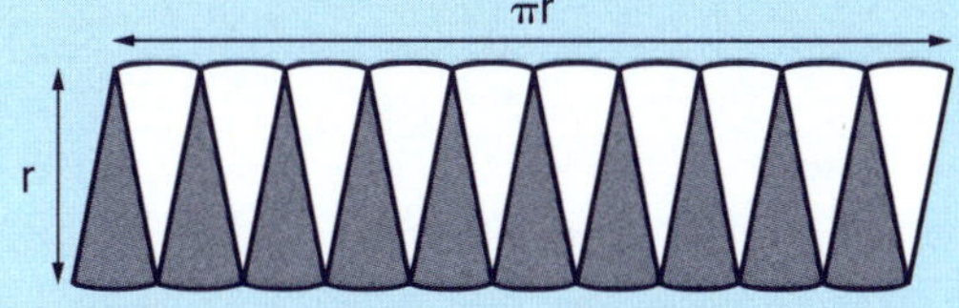

The formula for finding the area of a rectangle is $A = l \times w$.

The area of the sectors would therefore be $\pi r \times r$ (or πr^2).

The formula for finding the area of a circle is $A = \pi r^2$.

1 Follow these steps to show how circumference relates to finding the area of a circle.

- Use your compass to draw a circle with a radius of 8 cm onto a piece of paper.
- Cut out the circle.
- Fold the circle in half, then in half again and in half one more time.
- Open the paper up. It should be divided into eight equal sections.
- Cut along the fold lines and then reorganise the segments to form a rectangle.
- Measure the approximate (ignoring the curved edges) length and the width of your rectangle and use the formula $A = l \times w$ to calculate the rectangle's approximate area.

2 Calculate the area of a circle with an 8 cm radius by using the formula $A = \pi r^2$.

3 a Write a few sentences to explain why cutting sectors can only ever provide an estimation of the area of a circle.

b What would you expect to happen to your estimation using the sector method as you increase the number of sectors that you divide the original circle into? Why?

4 Test your suggestion by drawing another circle with an 8 cm radius. This time, cut the circle into four sectors only and re-form these into a rectangle. Measure the length and width to calculate the area of the rectangle. Discuss what you found with a partner.

5 Use the formula $A = \pi r^2$, where π represents 3.14, to find the area of the following circles.

a

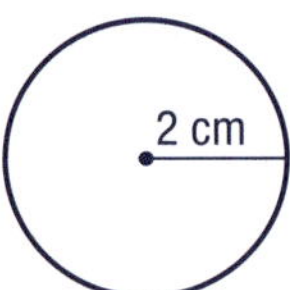

b

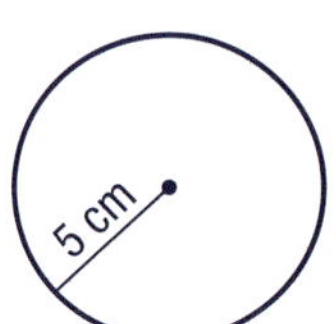

c

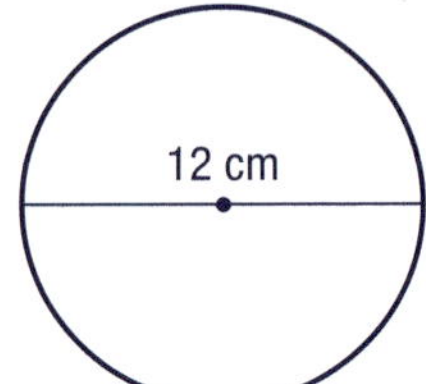

6 Using 3.14 to represent π, calculate the area of a circle with a radius of:

a 10 cm
b 20 cm
c 50 cm
d 1 m

Challenge

Calculate the area of a circle with a circumference of 31.4 cm.

Lesson 8 Calculating the area of a circle

The formula for finding the area of a circle is πr^2 where $\pi = 3.14$.

The symbol π can also be written as the fraction $\frac{22}{7}$.

Help Box

Calculate the area of a circle using the formula πr^2 using the two values for π given above.

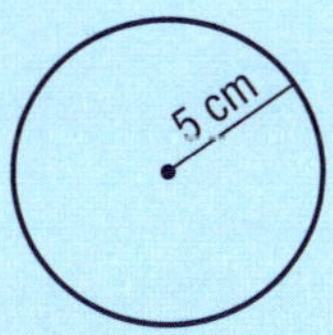

The radius of this circle is 5 cm.

Formula	$A = \pi r^2$	OR	$A = \pi r^2$
	$A = 3.14 \times (5 \times 5)$		$A = \frac{22}{7} \times (5 \times 5)$
	$A = 3.14 \times 25$		$A = \frac{22}{7} \times 25$
	$A = 78.5\text{ cm}^2$		$A = \frac{550}{7}$
			$A = 78\frac{4}{7}\text{ cm}^2$

If the diameter is given instead of the radius, remember to halve the diameter before applying the formula. The radius is half the length of the diameter.

1 Use π as either 3.14 or $\frac{22}{7}$ to calculate the area of the following circles.

a

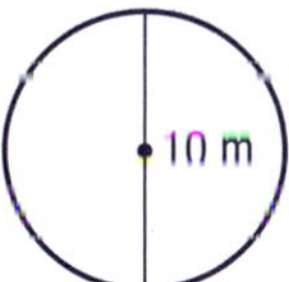

b

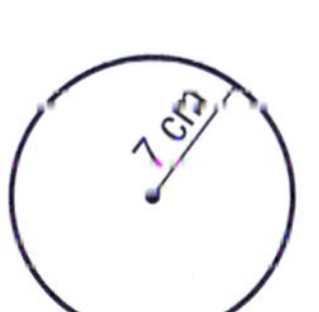

c

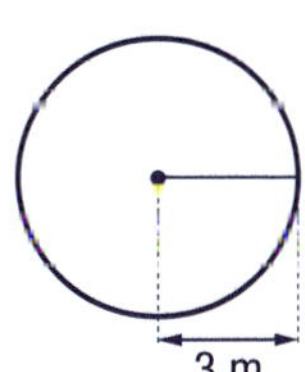

d

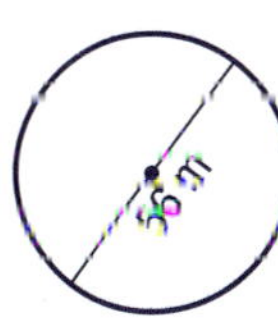

e

f

g

h

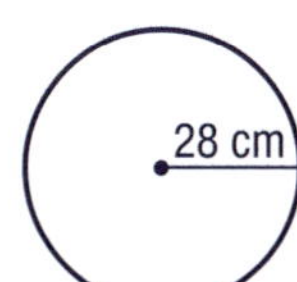

i

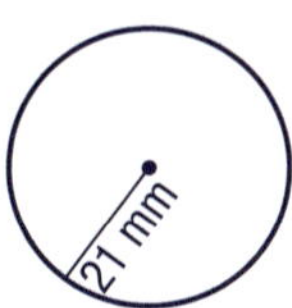

j

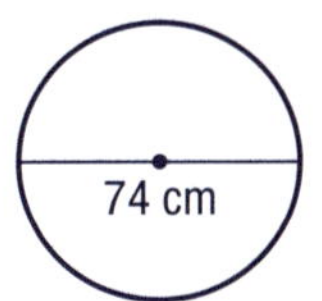

2 Find the area of the following shapes.

a

b
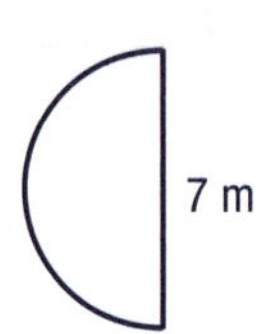

c
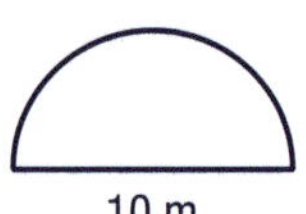

d

e

f
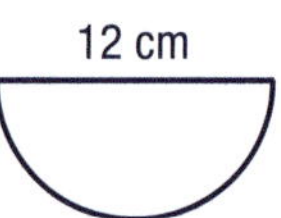

g

h
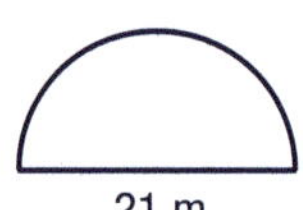

Challenge

Find the area of this shape.

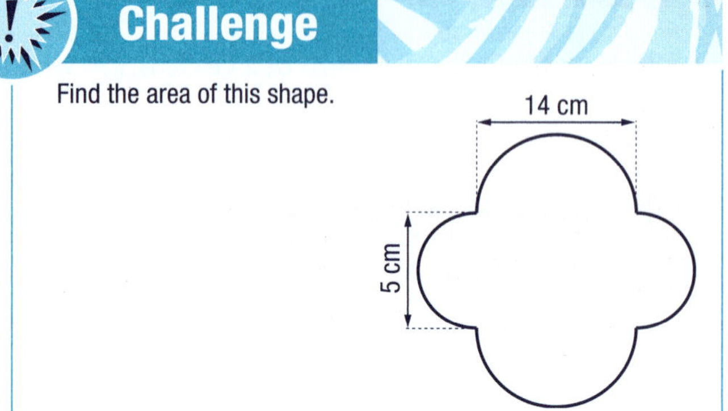

Lesson 9 Part circles

Some areas are a combination of circles and other shapes. They are called **compound shapes**.

1 Measure this compound shape and calculate:

a the perimeter
b the area

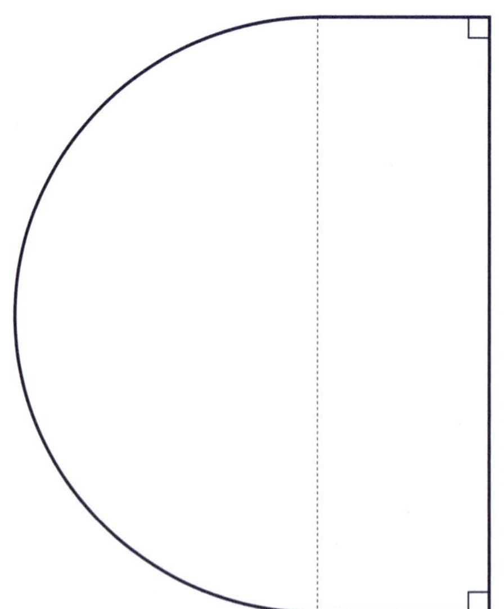

Help Box

A compound shape is comprised of two or more other shapes. Find the area of a compound shape by adding together the areas of the smaller shapes that it contains.

For example, this compound shape is comprised of two half circles and a rectangle.

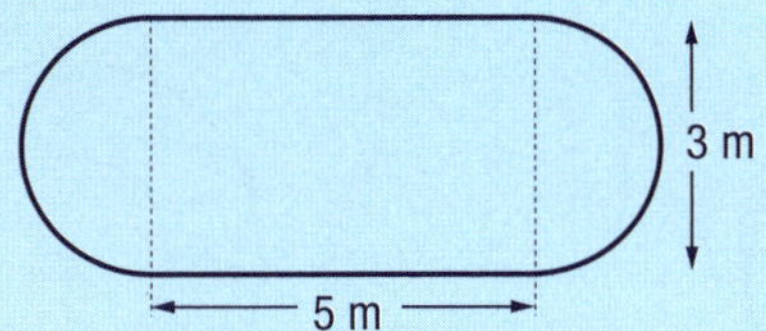

Area of rectangle

$A = l \times w$

$A = 5 \times 3$

$A = 15\ \text{m}^2$

Area of circle

$A = \pi r^2$

Find the radius ($\frac{d}{2}$ = radius), so $r = 1.5$

$A = 3.14 \times (1.5 \times 1.5)$

$A = 7.06\ \text{m}^2$

Area of compound shape = area of rectangle + area of circle

Area of compound shape = $15\ \text{m}^2 + 7.06\ \text{m}^2$

$= 22.06\ \text{m}^2$

2 Find the area of these compound shapes.

a

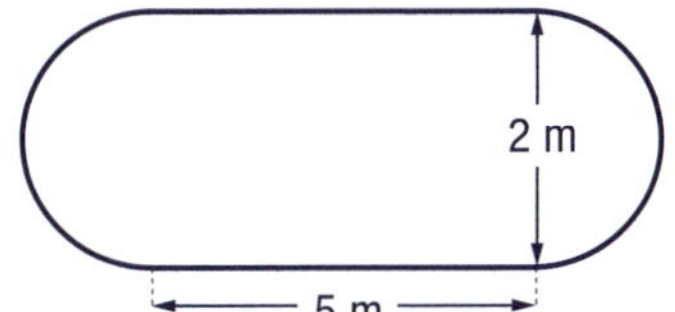

b

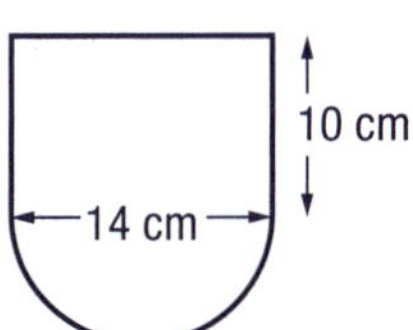

c

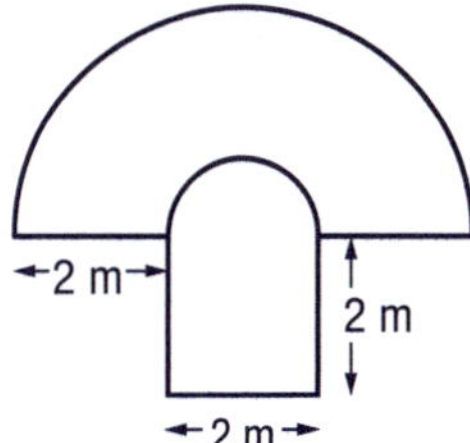

d

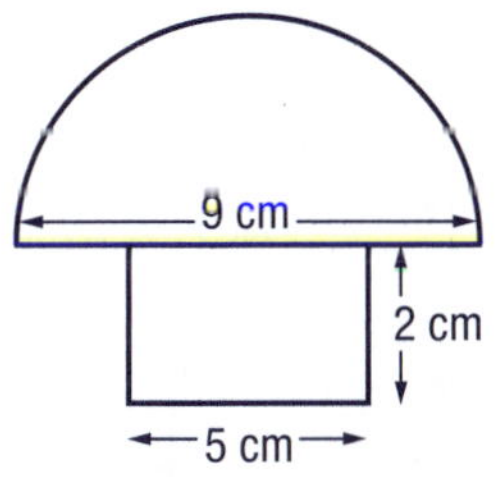

e

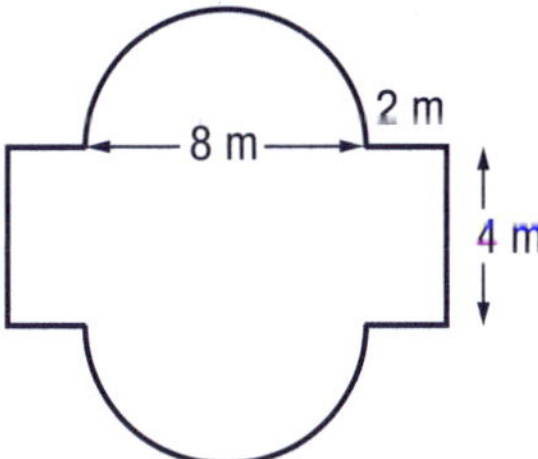

f

Help Box

Sometimes areas within circles need to be calculated.

For example, to find the area of the shaded region in this diagram we need to subtract the area of the inner circle from the total area of the larger circle.

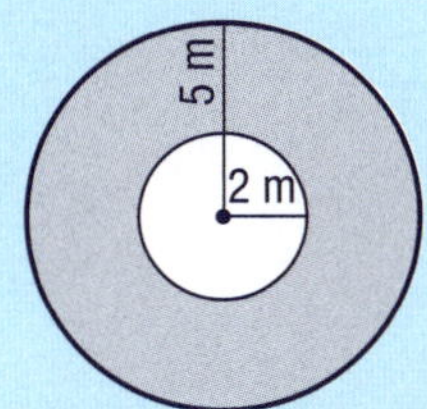

Area of outer circle

$A = \pi r^2$

$= 3.14 \times (5 \times 5)$

$= 78.5\ \text{m}^2$

Area of inner circle

$A = \pi r^2$

$= 3.14 \times (2 \times 2)$

$= 12.56\ \text{m}^2$ or $12.6\ \text{m}^2$

Area of the shaded region = area of the larger circle – area of the inner circle

Area of the shaded region = $78.5\ \text{m}^2 - 12.6\ \text{m}^2$

$= 65.9\ \text{m}^2$

3 Wesley has a small lawn mowing business. He charges on the basis of the total area to be cut. The shaded regions in the diagrams show the areas to be cut at each site. Estimate the area for each example, then calculate each area.

a
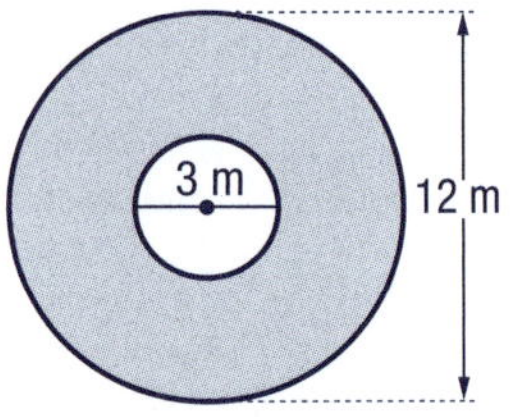

b
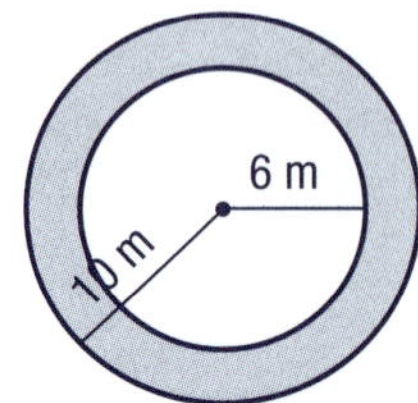

c
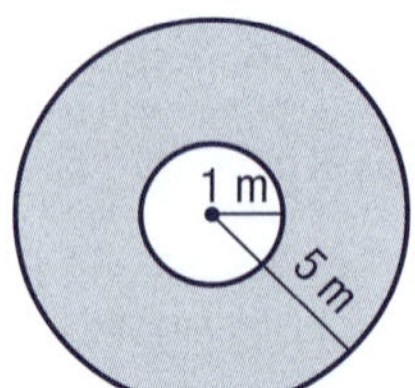

d
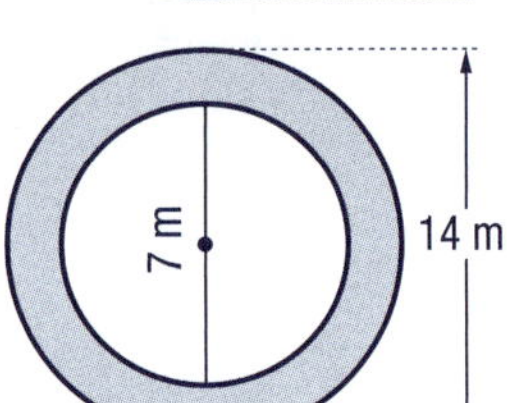

e
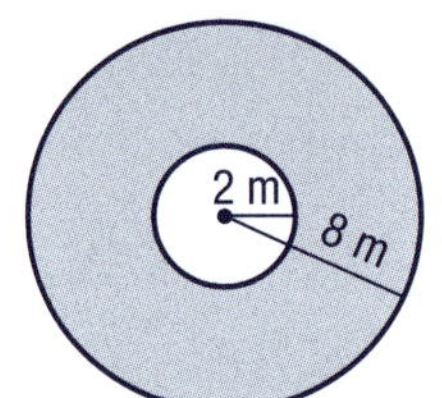

f
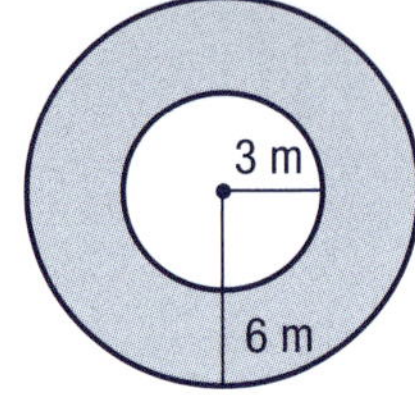

g
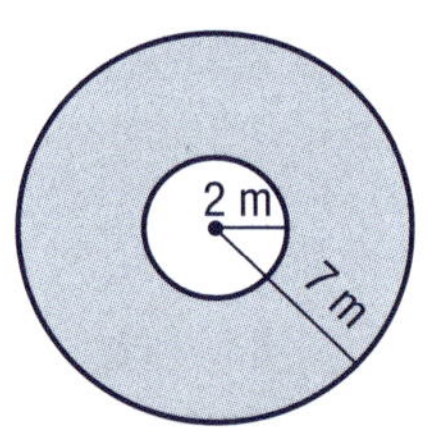

h
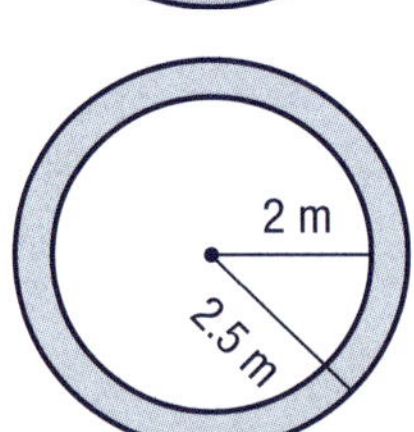

4 Find the area of the following shaded sections.

a
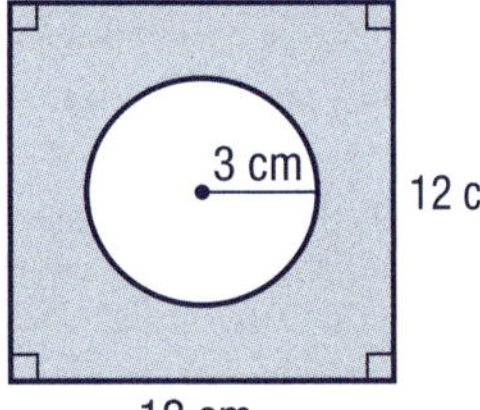

b
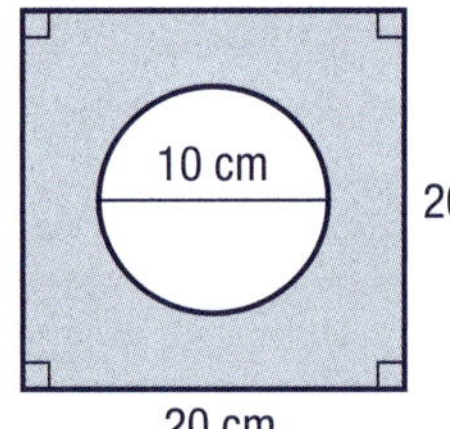

c
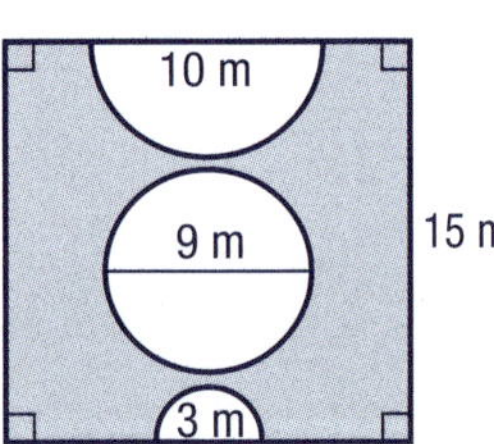

d
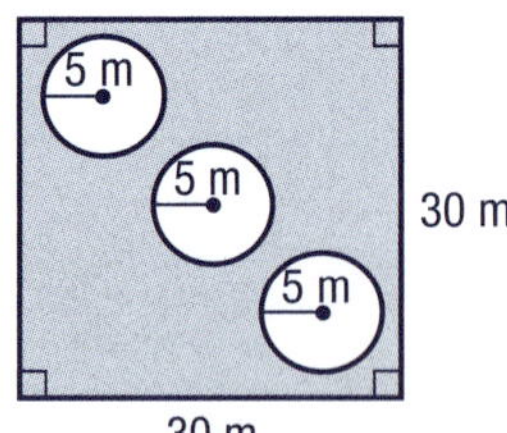

e
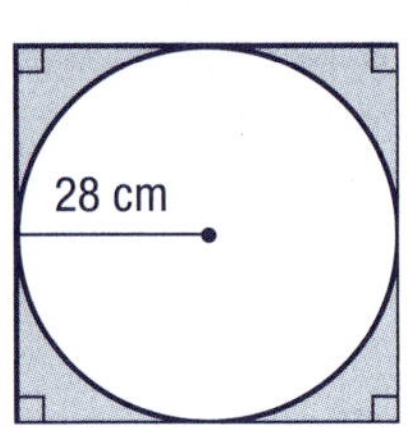

f
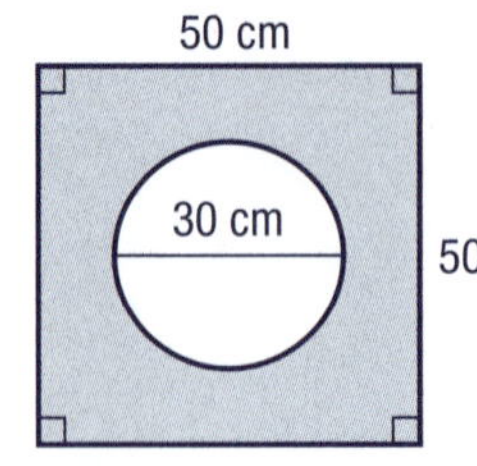

Challenge

Find the area of the shaded section.

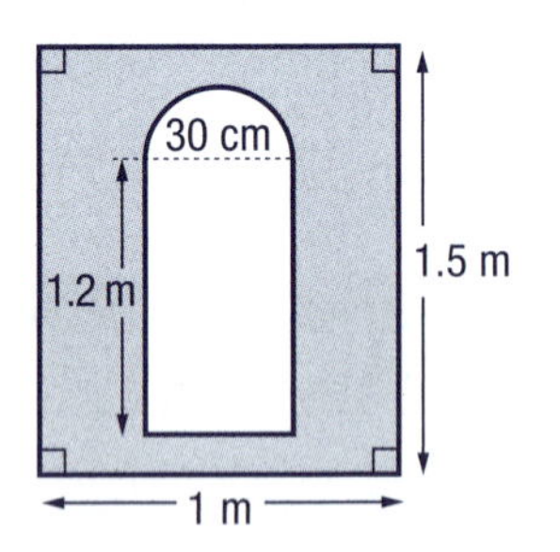

Lesson 10 Sets

You have previously learned how to use Venn diagrams to classify items into sets. This Venn diagram is labelled to help you remember some of the earlier work on sets that you have done.

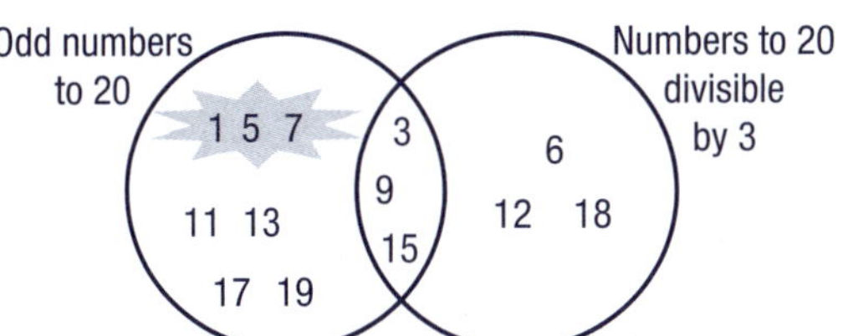

Elements in intersecting set (3, 9, 15)
Example of subset is odd numbers < 10 (1, 5, 7) that are not divisible by 3.

This Venn diagram compares some of the features of triangles and rectangles.

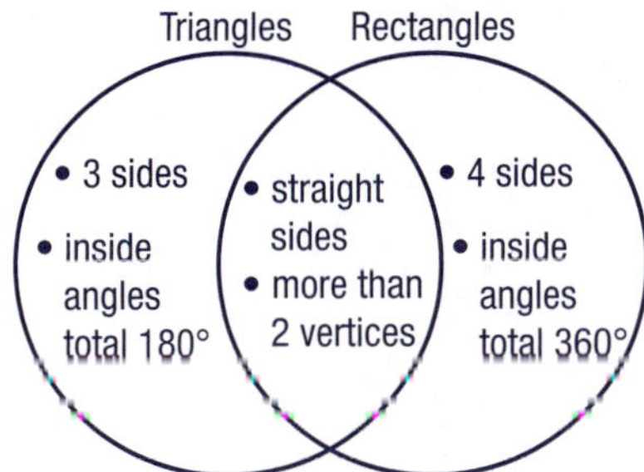

1 List all the items in the union of the two sets in the second Venn diagram.

2 List the set of elements from the Venn diagram that identify features of rectangles.

3 List the elements in the intersecting set between the features of rectangles and the features of triangles.

4 Describe a subset within one of the sets shown.

An **empty set** (sometimes called a **null set**) is a set that does not have any elements. For example, the set of shapes that are both circles and squares is a null set.

5 Describe another example of a null set.

6 Answer the questions relating to the Venn diagram.

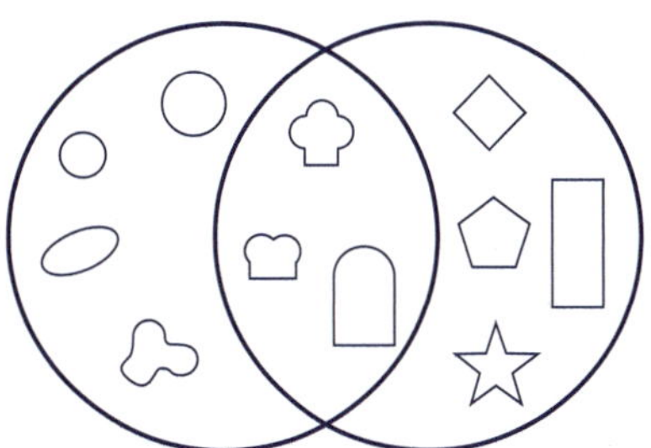

- a What could be the title of each of the two outer circles?
- b How many shapes are in the union of the two sets?
- c Draw the shapes in the intersecting set.

7 Wesley drew a Venn diagram to show the mowing work his company does.

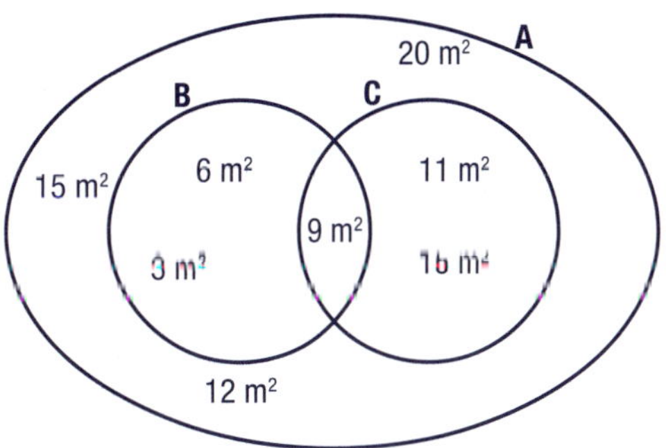

A: Set of jobs by grassed area (m^2)
B: Set of jobs less than 10 m^2
C: Set of jobs involving circular mowing

- a What is the total number of mowing jobs the company does?
- b How many of the company's jobs are less than 10 m^2 in area?
- c How many of those involve circular mowing?
- d List the size in m^2 of each non-circular job over 10 m^2.

8 A gardening company has 34 clients. If 17 require mowing, 16 require weeding and 5 only want watering, draw a Venn diagram to find out the number of clients who:

a only want mowing

b only want weeding

c want both mowing and weeding

d do not want weeding

9 In a village there are 33 families. Eight families have bicycles and 15 have wheelbarrows. There are 12 families who have neither a bicycle nor a wheelbarrow. Draw a Venn diagram to find out the number of people in the village who:

a only own a bicycle

b only own a wheelbarrow

c own both a bicycle and a wheelbarrow

d do not own a bicycle

10 Use the following descriptions to collect information from students in your class and then draw Venn diagrams to show what you found. Label each set appropriately and be sure to include the set of students who do not fit into either of the described sets.

a The set of students who own a bicycle and the set of students who own a wheelbarrow.

b The set of students who know how to use a sewing machine and the set of students who know how to use a chainsaw.

c The set of students who like to be outdoors and the set of students who like working with machinery.

11 a Select one of the Venn diagrams you drew in Question 10 and write three questions that would test someone's ability to interpret the Venn diagram.

b Write the answers to your questions.

c Swap your diagram and questions with a friend and ask them to work out the answers.

Challenge

Complete the Venn diagram by filling in the missing numbers. Label each set with an appropriate title and write five statements about the information presented to show how well you understand Venn diagrams.

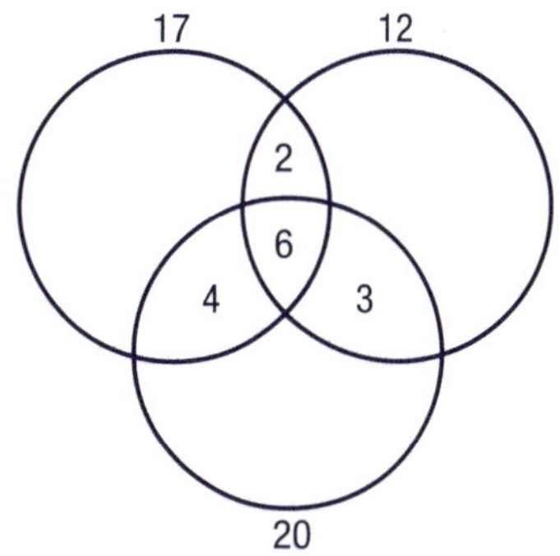

Learning Unit 3 In the Tool Shed

Strand: Number and Application

Fractions	Outcome 8.1.1	Apply fractions in problem solving
Fractions and Decimals	Outcome 8.1.3	Convert freely between fractions, decimals, percentages and ratios
Ratios and Rates	Outcome 8.1.5	Apply ratios in solving problems from real life

Strand: Space and Shape

Angles and Shape	Outcome 8.2.11	Investigate properties of interior and exterior angles of polygons
Angles	Outcome 8.2.12	Use appropriate terms to describe angles and shapes accurately

Strand: Chance and Data

Estimation	Outcome 8.4.7	Estimate results of calculations

Lesson 1: Introduction	Looking at simple tools and their uses Making some simple tools for drawing circles and ellipses
Lesson 2: Finding angles	Measuring angles without a protractor Constructing set squares
Lesson 3: Measuring and calculating angles	Using the mitre box Calculating interior and exterior angles
Lesson 4: Problem solving involving angles	Investigating adjacent, complementary, supplementary and vertically opposite angles
Lesson 5: Parallel lines are special	Investigating corresponding, alternate and co-interior angles
Lesson 6: Polygons and angles	Calculating interior and exterior angles of polygons
Lesson 7: Fractions and percentages	Solving problems
Lesson 8: Comparing common fractions and decimals	Exploring spanners and international measurement standards Converting between decimals and fractions
Lesson 9: Ratio	Applying ratio to gears Relating ratio and fractions
Lesson 10: Problems in the tool shed	Using ratios in real-life situations Linking fractions, decimals, percentages and ratios

Lesson 1 Introduction

Tool sheds can be a wonderful source of inspiration for people who are prepared to use their imagination and creativity. Building and repairing your own furniture and equipment can be a very good way for a family to save money. Producing extra items to sell to friends or at the market is a way of increasing the family income.

During this unit we will examine some of the simple tools that people use to make difficult tasks easier. All tools involve mathematics to some degree. Some of the tools to be examined are commercially made while others are simple tools that people can build or create themselves from scrap material.

1 What are the names of these simple tools?

2 What are they used for?

3 How else could that task be completed if a person did not have this tool?

4 Name some other tools that your family uses to make certain tasks easier.

5 Of all the tools owned by your family, which do you think is the most important? Why?

6 Make these two simple tools: one to help you draw a circle when you don't have a compass and the other to help you draw an ellipse.

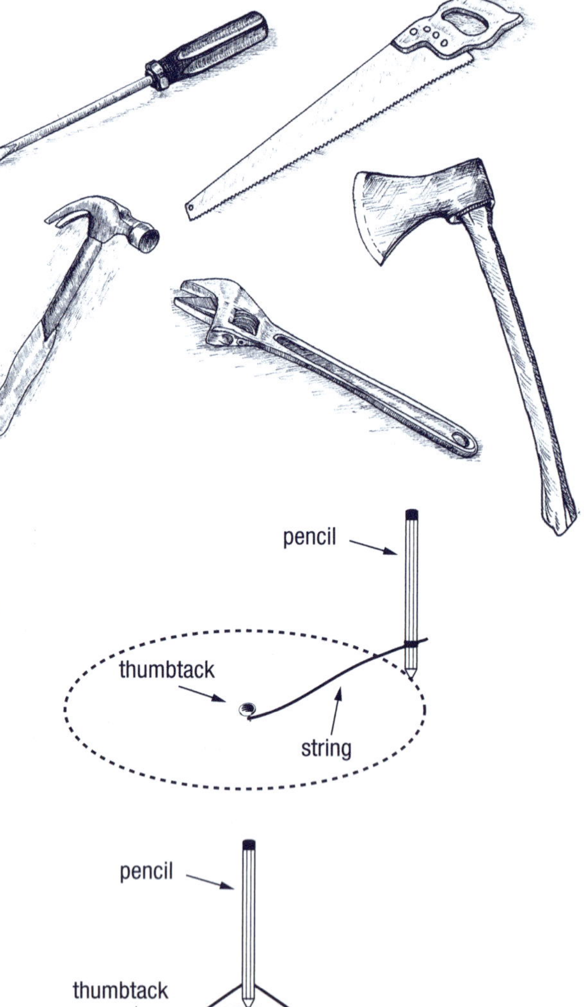

TOOL 1: Tie a knot in the end of a piece of string and tie the other end around a sharpened pencil. Push a thumbtack through the knot and hold it firmly with your thumb. Stretch the string taut and, holding the pencil upright, draw a circle around the thumbtack.

TOOL 2: Tie a thumbtack to either end of a piece of string. Push the thumbtacks into a piece of cardboard leaving some slack in the string. Draw an ellipse by sliding a pencil along the string each side of the line between the two tacks.

Predict how the shape would change if the thumbtacks were placed closer together or further apart. Test your predictions.

Lesson 2 Finding angles

Many building and craft projects require us to first identify a right angle (90°). Folding a piece of paper twice can create a right angle. Roughly tear a piece from a sheet of newspaper. Fold it once to make a straight angle and then once again to form a right angle. Open the paper out again and examine the crease lines to answer the following questions.

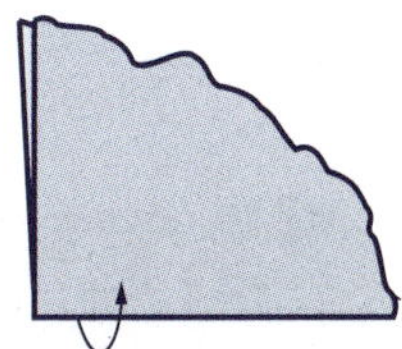

1 a How many right angles are there in a straight angle?

b How many right angles in a full revolution (360°)?

c If the paper had been folded a third time, what angle would have been formed between the new crease lines?

In Grade 7 you tore the corners from a triangle and rearranged them to form a straight line to prove that the sum of the internal angles of any triangle is 180°.

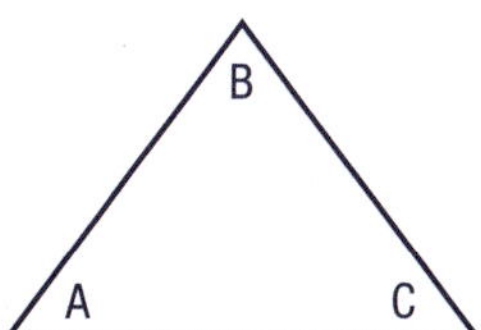

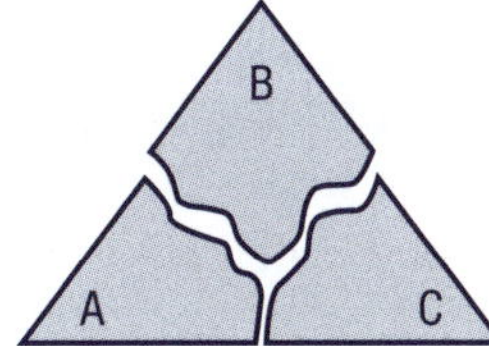

You did a similar activity with a quadrilateral to prove that the sum of the internal angles of any quadrilateral is 360°.

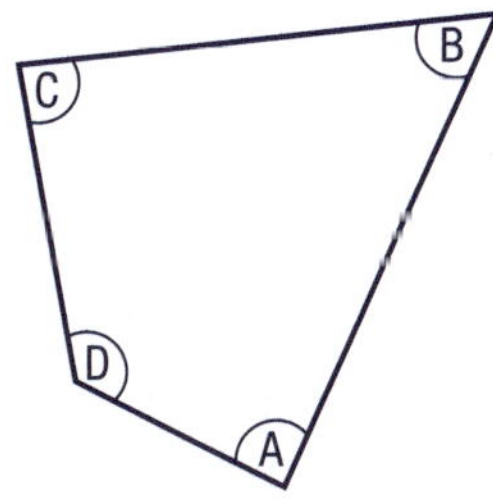

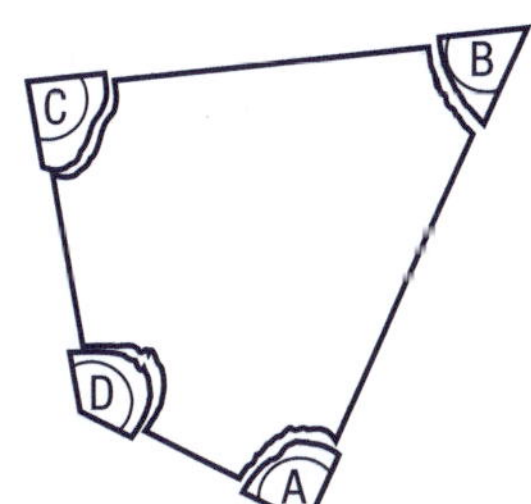

2 Find the missing angles in the following shapes.

a

b

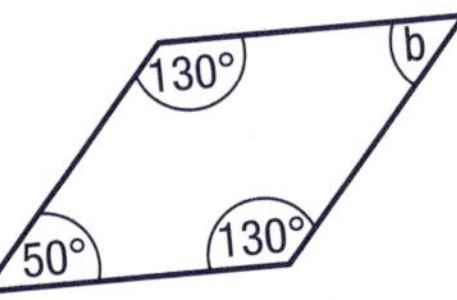

c

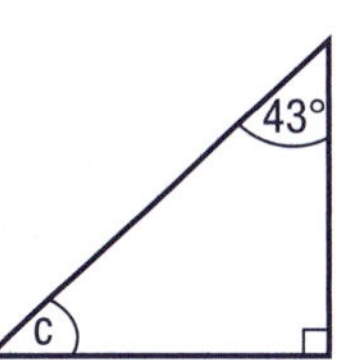

d

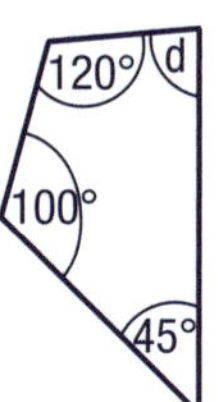

e

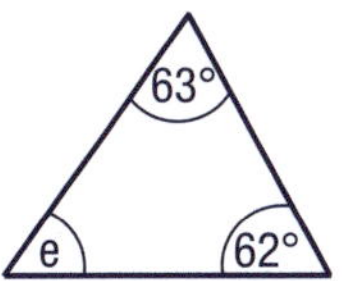

f

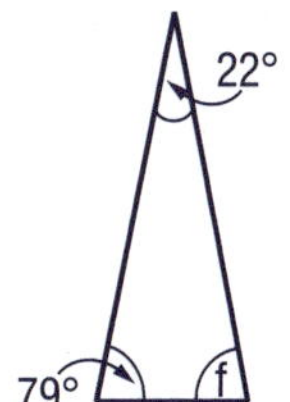

g

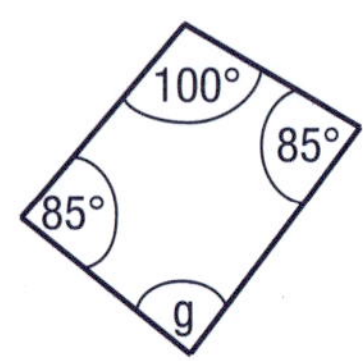

h

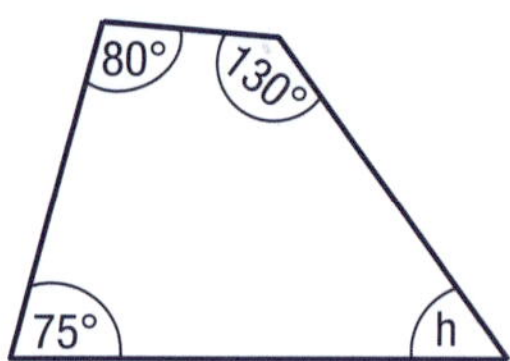

i

75°

i

Help Box

A triangle has a right angle if the square of the longest side is equal to the sum of the squares of the other two shorter sides. This is called the *rule of Pythagoras*.

The square of the side *XY* is $(5^2) = 25$ cm^2.

The sum of the squares on the other two sides *ZX* and *YZ* is $(3^2 + 4^2) = 25$ cm^2.

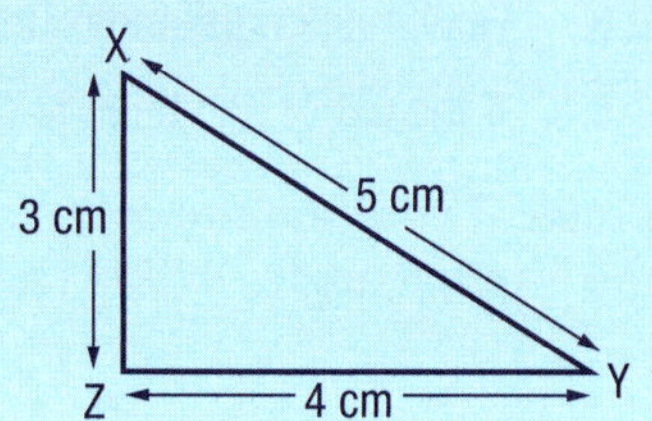

Thousands of years ago the Egyptians used the knowledge that a triangle with sides 3:4:5 is a right-angled triangle. It can be used to make a simple tool for forming right-angles in buildings and constructions.

3 a Work with some classmates to make and trial this Egyptian tool.

Step 1: Cut a piece of string (preferably over 50 cm long for ease of handling).

Step 2: Tie 12 knots equally spaced along the string.

Step 3: Form the string into a right-angled triangle as shown.

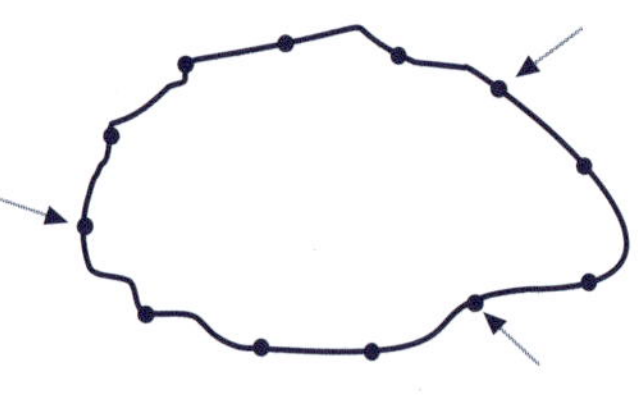

take hold of knots at arrows

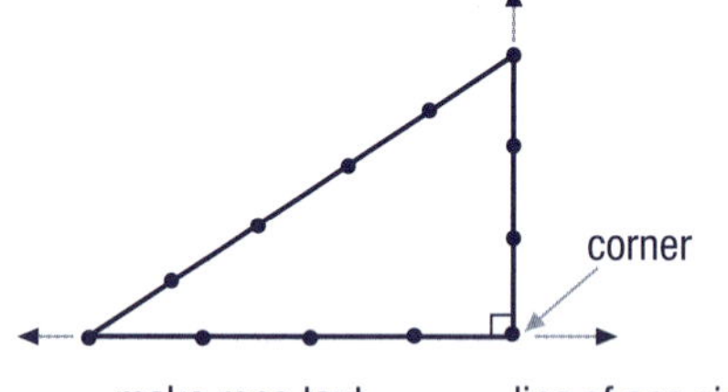

make rope taut

line of one side of building

b Experiment to see what happens when one side of the string triangle has a row of knots other than 3, 4 or 5. Write a few sentences explaining what you found.

A set square is a triangle that is used as a template to mark and draw particular angles onto paper, wood, metal or other materials. Set squares can be made from a sheet of stiff cardboard. The most common angles found on the corners of set squares are 22.5°, 30°, 45°, 60° and 90°. These angles can be formed very easily without a protractor by carefully measuring the length of the sides of a triangle.

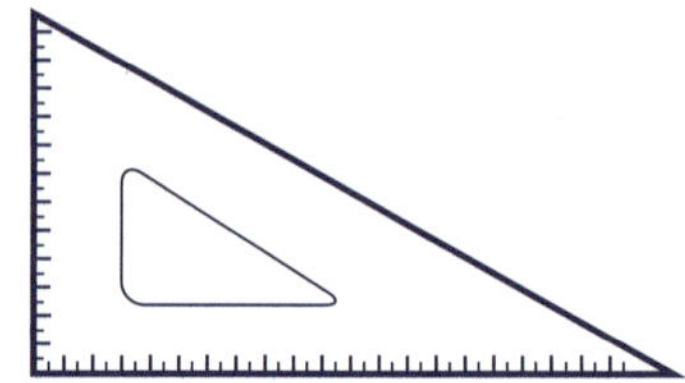

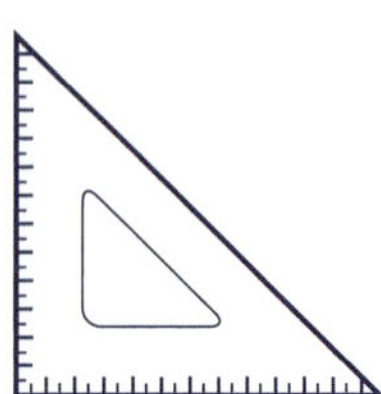

4 Measure and cut a piece of cardboard 30 cm by 21 cm (standard A4 size). Use this one sheet of cardboard for questions 5 and 7. Follow the instructions to check that all four corners of your piece of card are right angles.

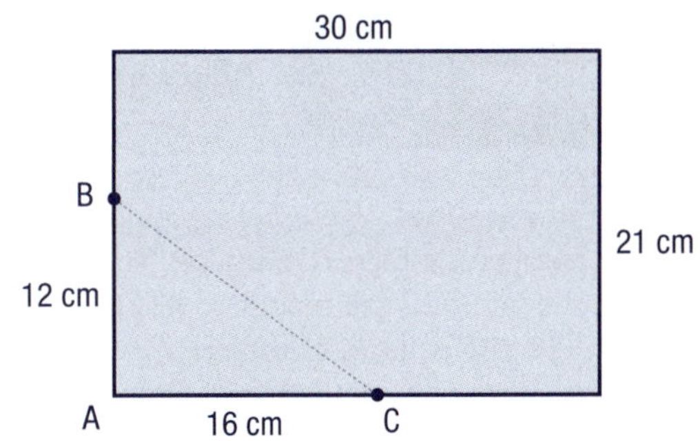

Step 1: Measure and mark a point *B* that is 12 cm from point *A*.

Step 2: Measure and mark a point *C* that is 16 cm from point *A*.

Step 3: If the side *BC* measures 20 cm, then the corner must be 90° (because 12:16:20 is a multiple of 3:4:5).

Challenge

List other sets of three measurements that could be used to check that the four corners are right angles. Explain how you worked out your answers.

5 Follow these instructions to make a 45° set square.

Step 1: From one corner, measure and mark 20 cm along both edges as shown in the diagram.

Step 2: Rule a diagonal line to join *P* to *Q*.

Step 3: Cut along the line *PQ*. The shaded area is the required set square.

Step 4: Label the corners of the set square with the angle sizes.

P
20 cm
A
20 cm
Q

6 Write a few sentences in your workbook to explain how you know that 45° angles can be drawn using the set square.

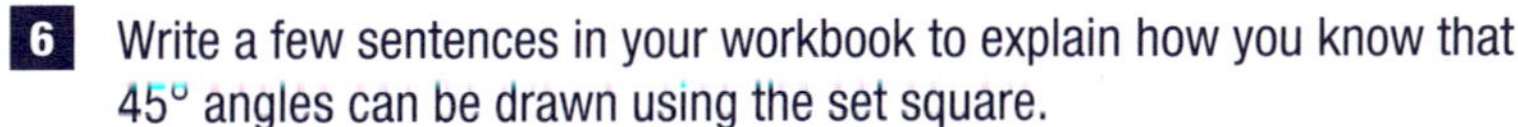

7 Follow the instructions to make a 30° and 60° set square from the remaining piece of cardboard.

Step 1: From the corner marked *X*, measure and mark 9 cm as shown in the diagram.

Step 2: Open a compass to 18 cm and, placing the point at *R*, draw an arc to locate *S* on the other edge.

Step 3: Rule a diagonal line from *R* to *S*.

Step 4: Cut along the line *RS*. The shaded area is the 30°, 60° set square.

Step 5: Label the corners of the set square with the angle sizes.

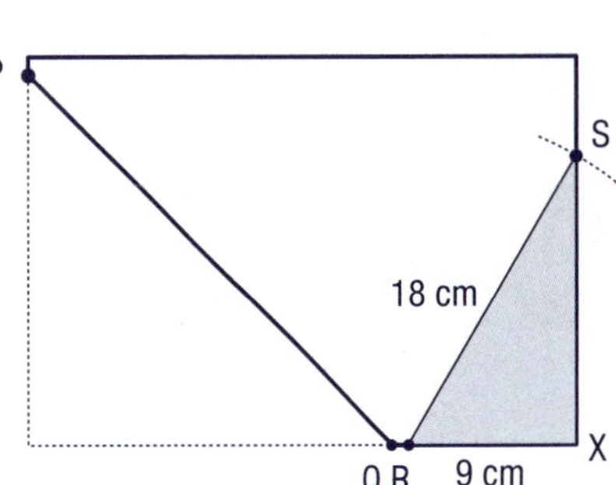

Lesson 3 Measuring and calculating angles

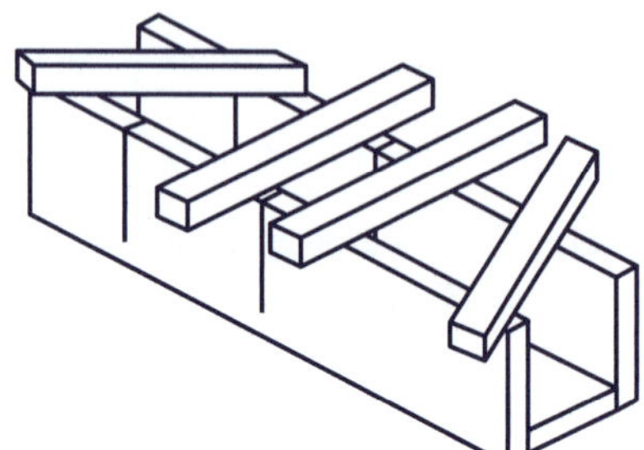

A mitre box is useful when sawing angles such as the 45° angles needed to make a rectangular picture frame. A three-sided mitre box wide enough to hold the width of the boards to be cut can be quickly made from scrap timber. The three pieces of timber are nailed together, and the top steadying pieces are fixed with smaller pins. Slots are cut into the walls of the box at the precise angle at which the cut is to be made. These slots help to guide the saw. The board to be cut is placed in the box, and the point where the board is to be cut is lined up with the appropriate slot in the mitre box wall.

1 Use a protractor to measure the angle that these boards have been cut at.

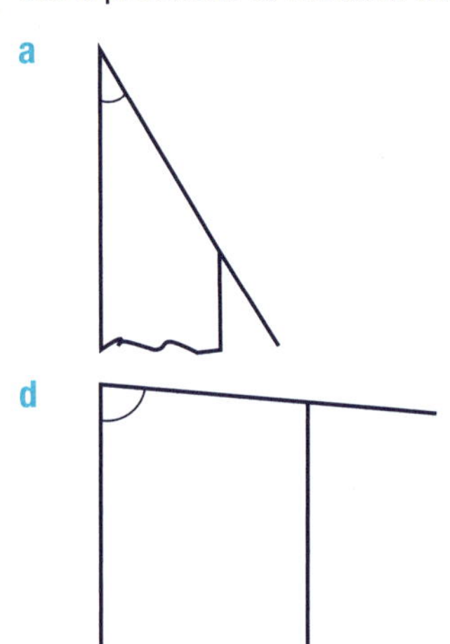

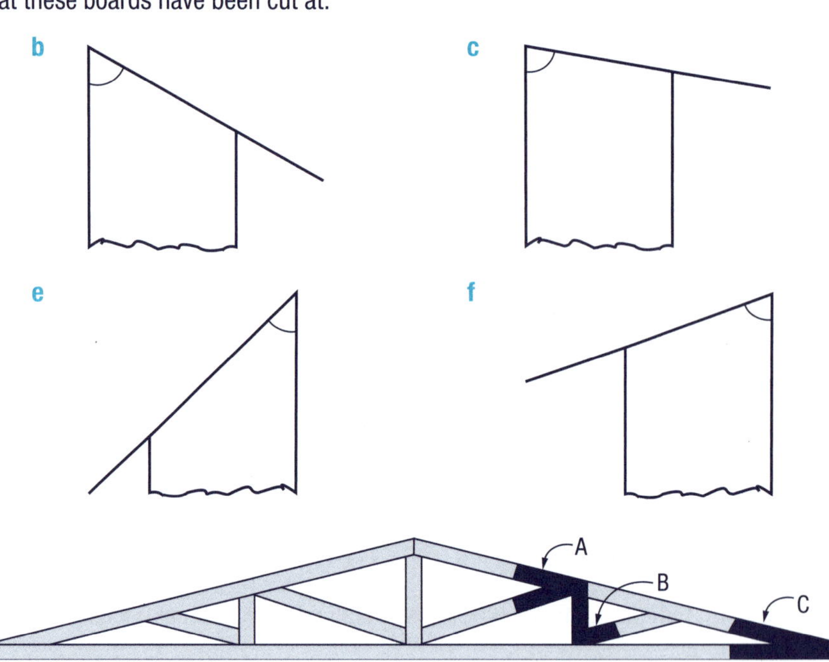

2 The framework used to support a roof is built using many angled pieces of wood. Use a protractor to measure the three different angles marked on the diagram to the right.

Help Box

There are 360° in a complete turn.

There are 180° in a straight line.

There are 90° in a right angle.

The sum of the *interior angles* of a triangle is 180°.

The sum of the *interior angles* of a quadrilateral is 360°.

(Interior angles are sometimes called *internal angles.*)

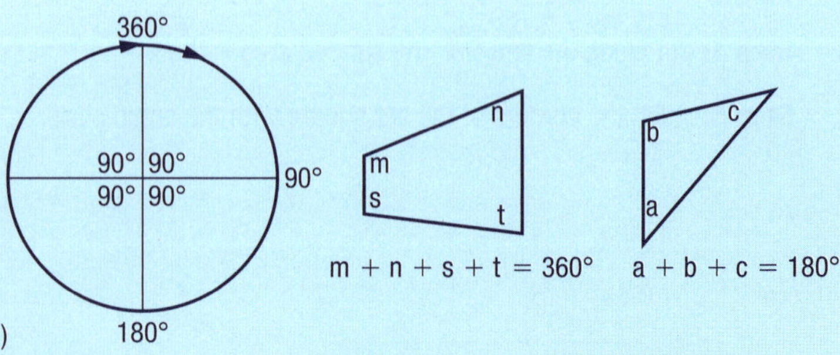

3 Use your knowledge of angles to find the missing interior angles in the following examples.

a

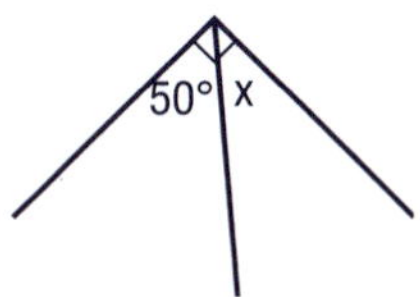

b

c

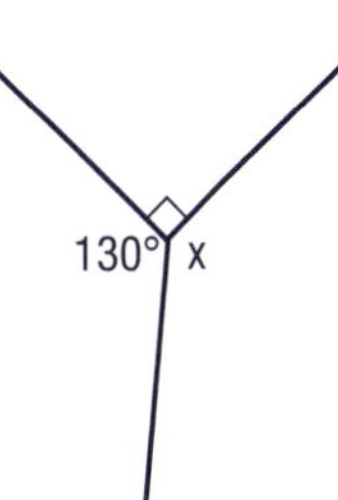

d

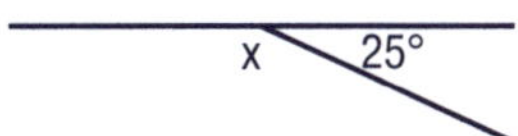

e

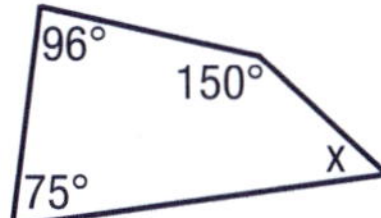

f

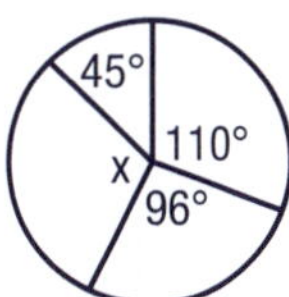

g

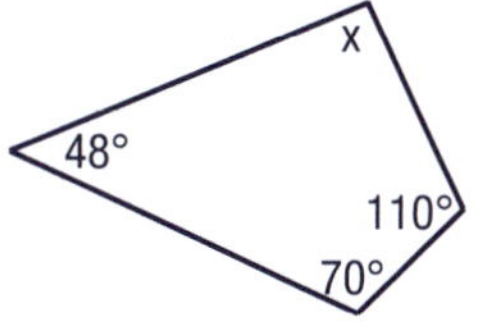

Help Box

The *exterior angle* is the angle outside a shape. The exterior angle of this triangle is between a side that has been extended and another side as shown.

(*Exterior angles* are sometimes called *external angles*.)

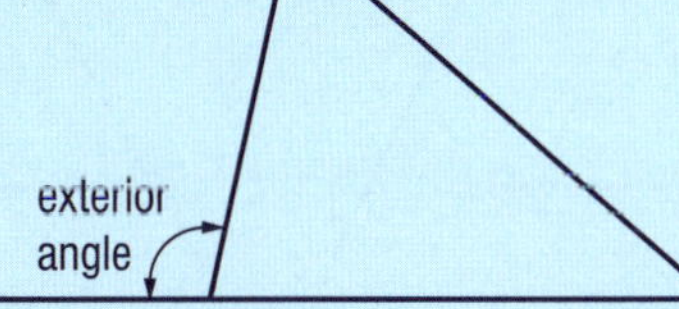

4 Use your knowledge of angles to explain how you could work out the exterior angle (a°) in the following diagram without using a protractor.

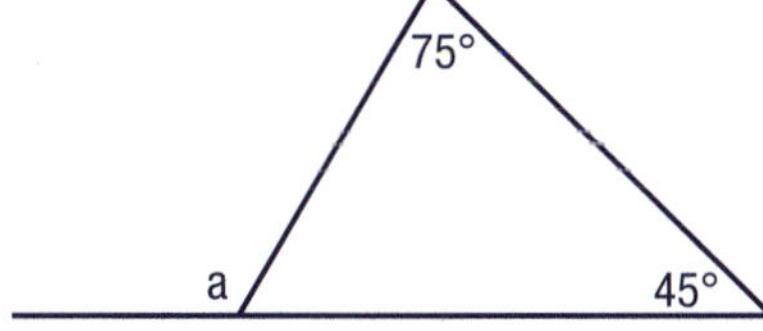

5 Find the external angle in each example without using a protractor.

a

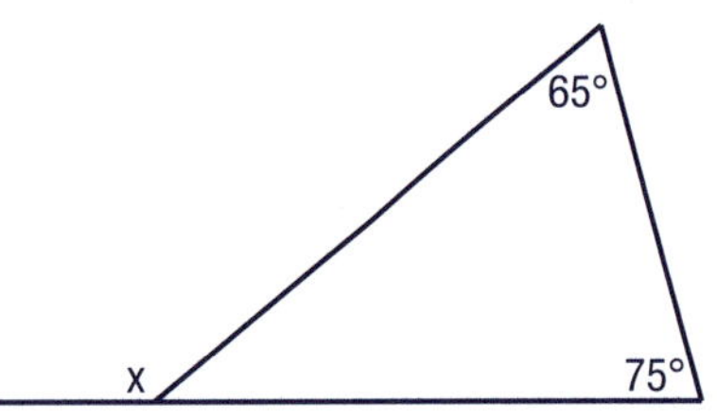

b

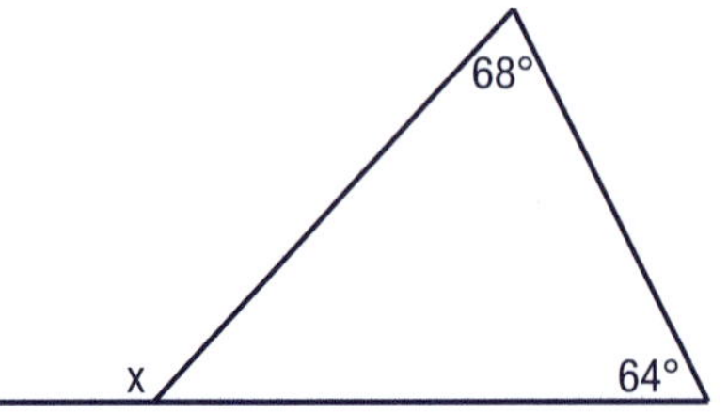

c

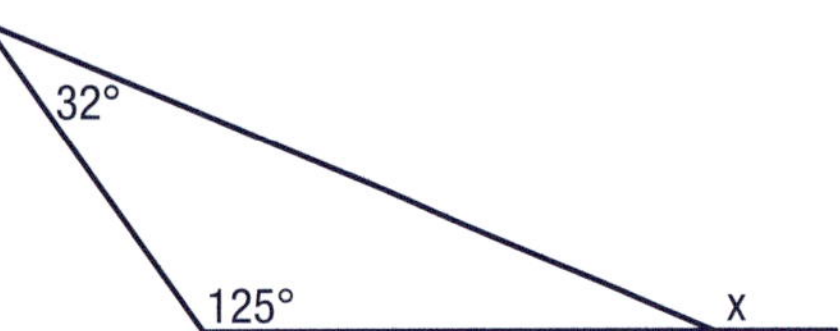

d

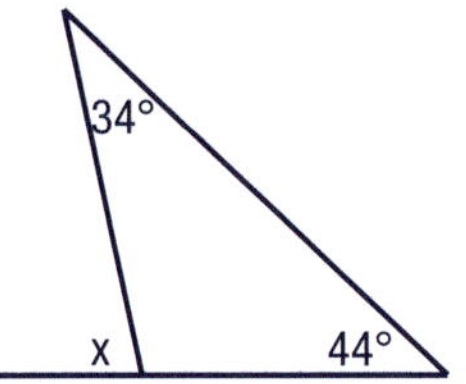

e

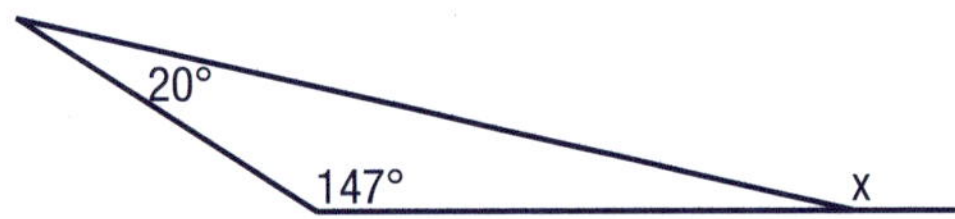

f

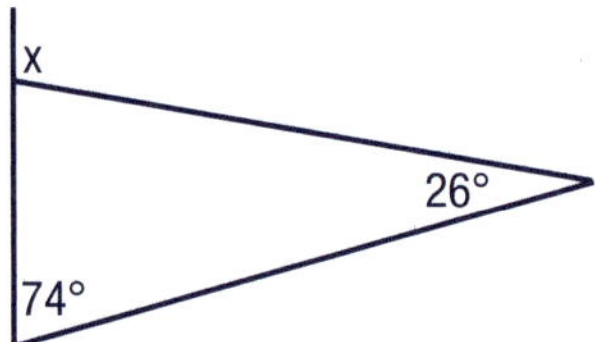

g

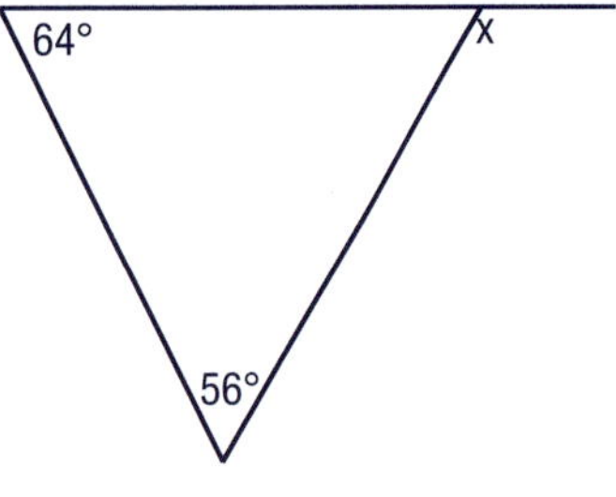

h

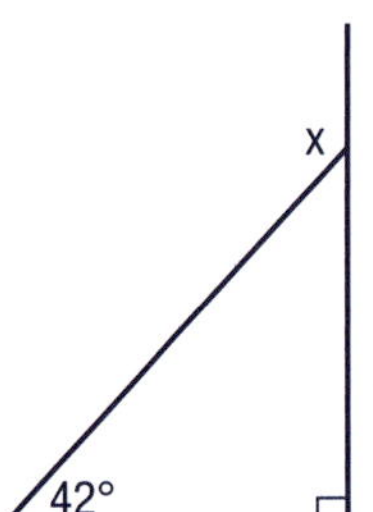

i

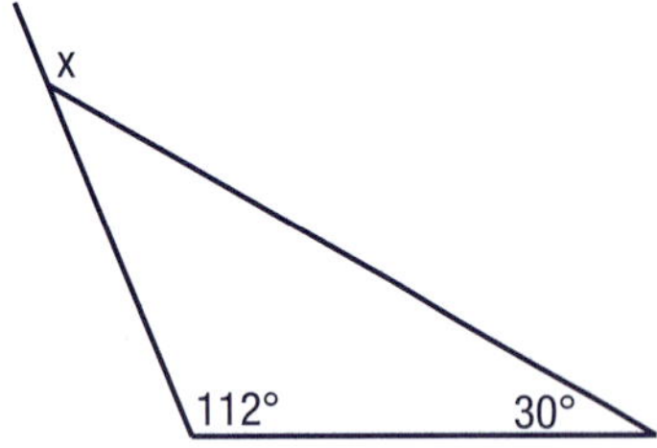

Challenge

Investigate whether it is possible to predict the sum of the exterior angles of any triangle or quadrilateral.

Lesson 4 Problem solving involving angles

In the last lesson you used your knowledge of angles to calculate some unknown angles. This lesson will build on that understanding.

1 In the diagram $\angle ABC$ is adjacent to $\angle CBD$.

a Describe two other adjacent angles in the diagram.

b Describe two angles that are not adjacent.

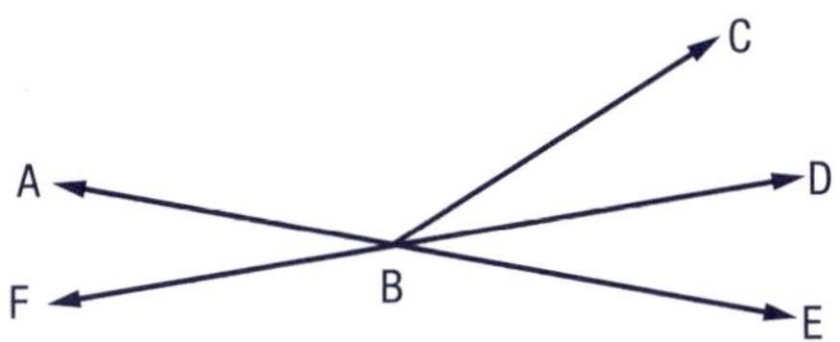

Help Box

Adjacent angles share a common ray and have the same vertex.

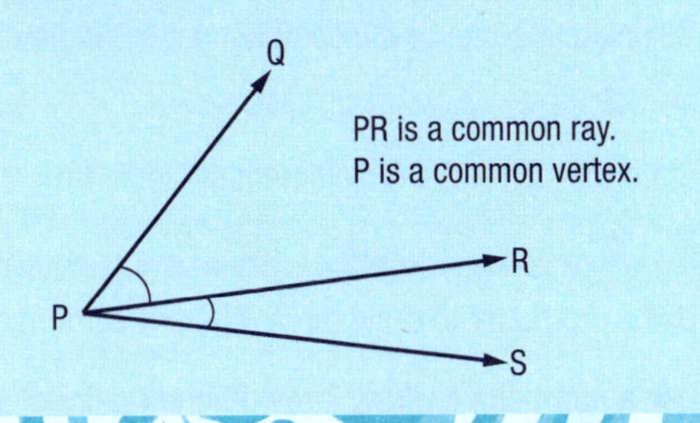

Help Box

When two adjacent angles add to 90° they are called **complementary angles**. When two adjacent angles add to 180° they are called **supplementary angles**.

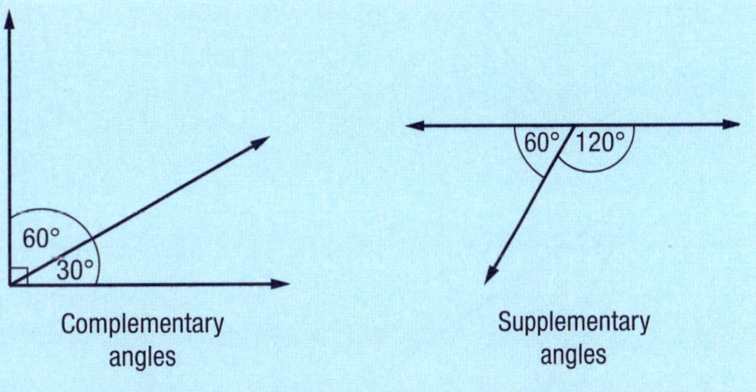

2 Calculate the size of the missing angle in each of the following examples. State whether the two adjacent angles are complementary or supplementary.

a

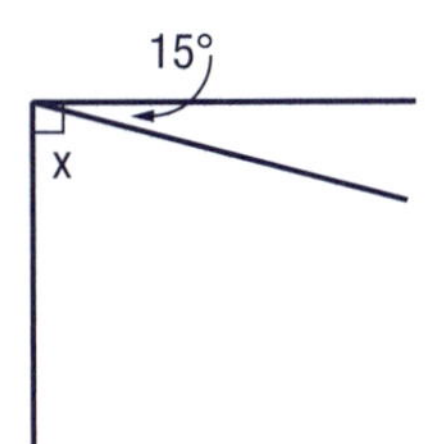

b

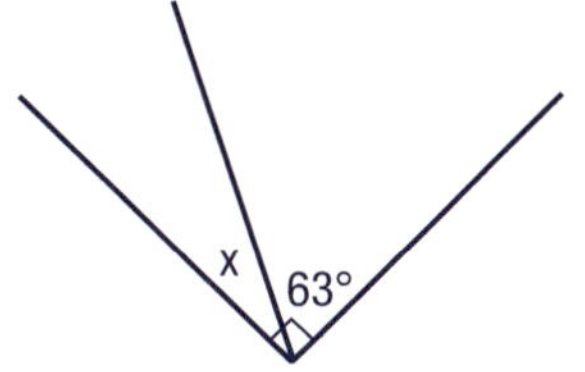

c

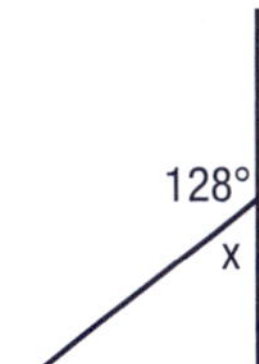

d
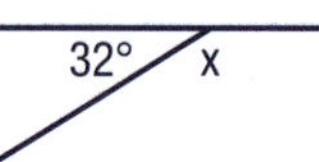

e
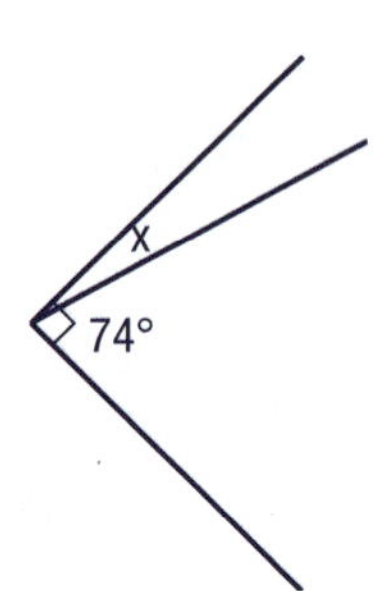

f
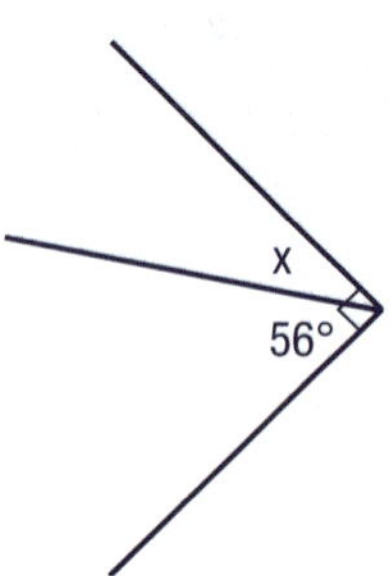

3 Calculate the size of the complementary angle to:

a 5° b 19° c 37° d 56° e 74°

4 State the size of the supplementary angle to:

a 9° b 68° c 85° d 107° e 149°

5 Use a protractor to draw three different pairs of complementary angles. Write the sizes of the angles on your diagrams.

6 Use a protractor to draw three different pairs of supplementary angles. Write the sizes of the angles on your diagrams.

7 Calculate the value of the unknown angles *y* in each of the following examples without using your protractor.

a
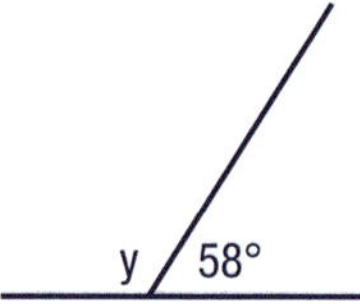

b
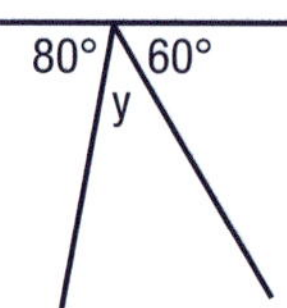

c
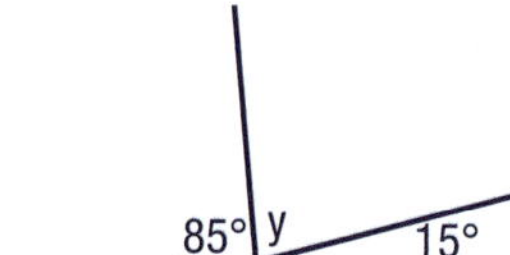

d
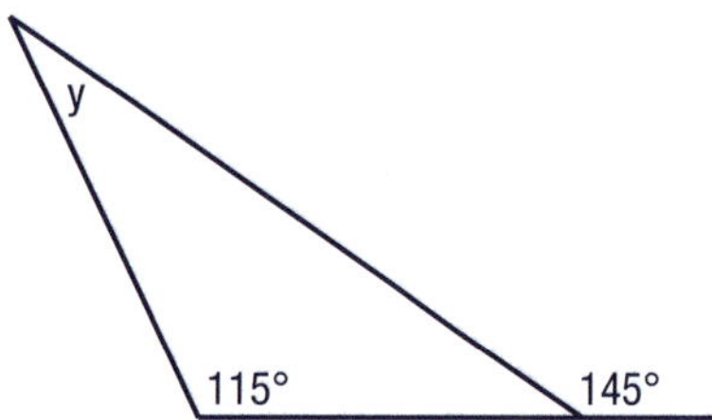

e
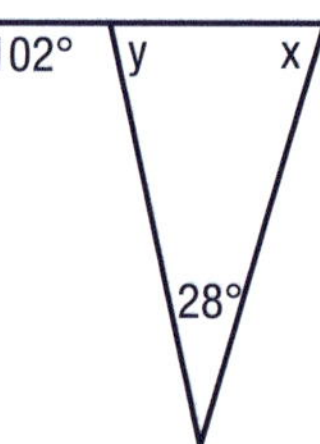

8 a Use your ruler to draw two 5 cm straight lines that cross each other.

b Using a protractor, measure and label the angles in your diagram.

c Discuss with a friend any relationships you notice between the angles.

Help Box

When two straight lines cross each other, two pairs of equal supplementary angles are formed. The vertically opposite angles are equal.

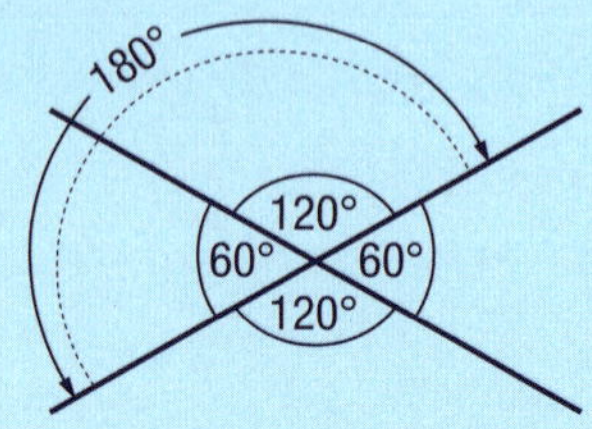

9 Calculate the missing angles without using a protractor.

a

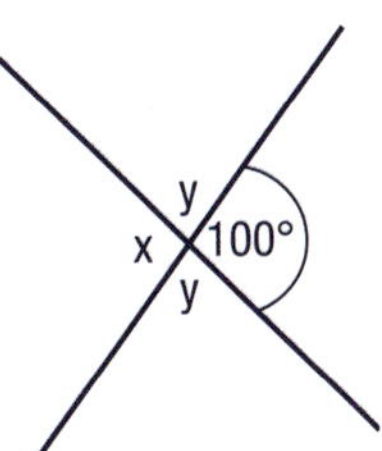

b

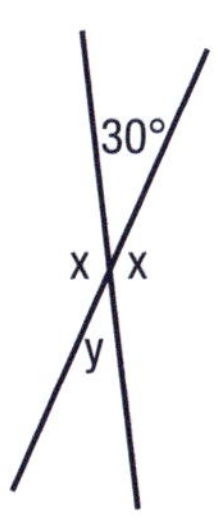

c

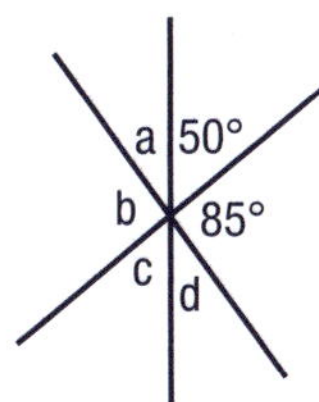

d

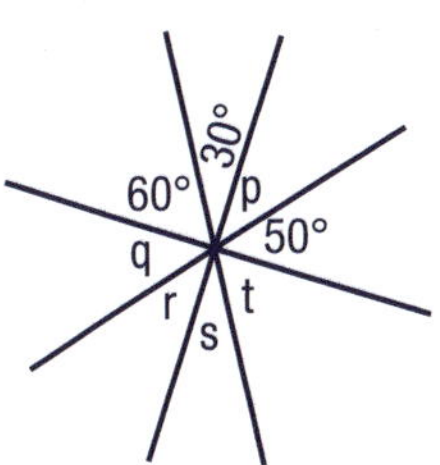

Lesson 5 Parallel lines are special

Most building materials have opposite sides that are **parallel**. A piece of timber is usually rectangular or square, and each edge is parallel to the opposite edge.

Help Box

Parallel lines are straight lines that are always the same distance apart. This diagram shows the arrow head symbol used in mathematics to indicate when lines are parallel.

A straight line that crosses (or transverses) a set of parallel lines is called a **transversal**. A transversal crosses both parallel lines at the same angle.

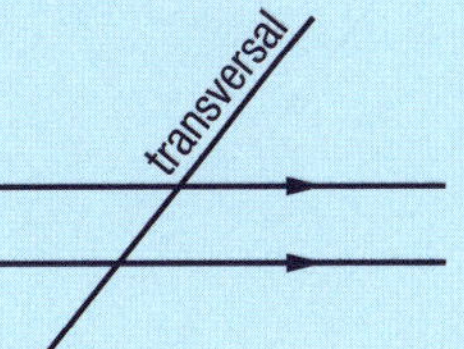

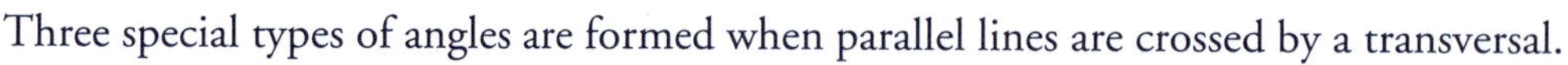

Three special types of angles are formed when parallel lines are crossed by a transversal.

- **Corresponding angles** are both above or both below parallel lines and are located on the same side of the transversal.

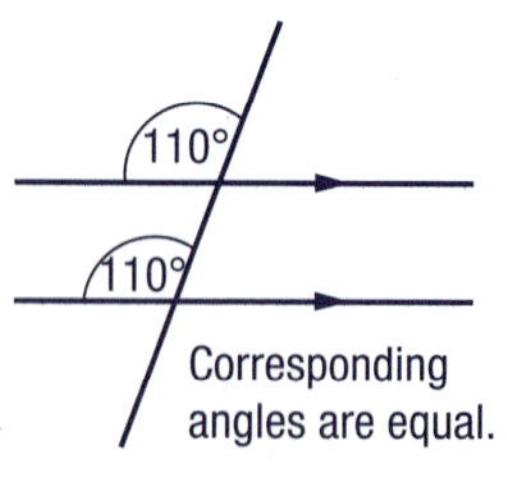

- **Alternate angles** are both between the parallel lines and on opposite sides of the transversal.

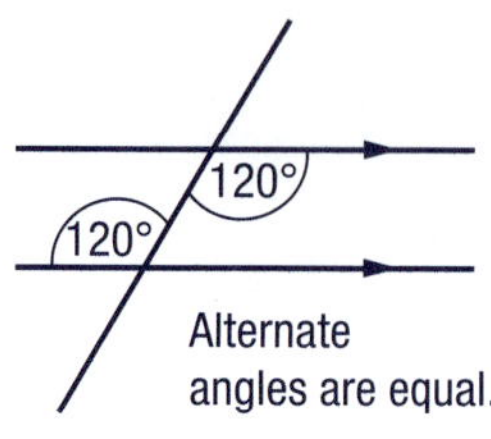

- **Co-interior angles** are also both between the parallel lines but are on the same side of the tranversal.

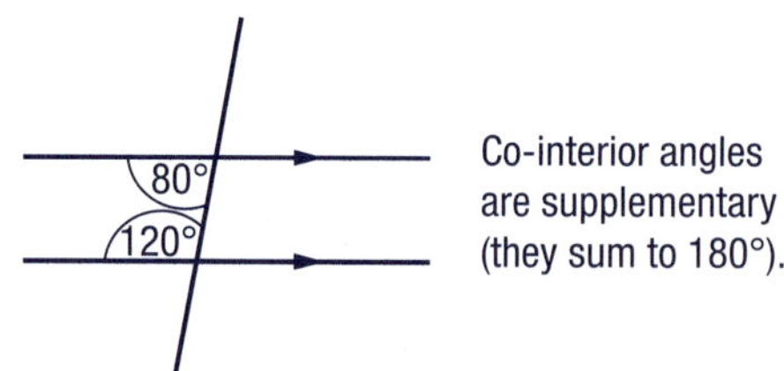

1 Which angle is corresponding to the angle marked *A*?

a

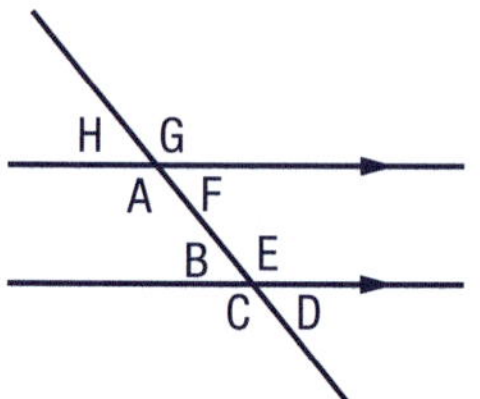

b

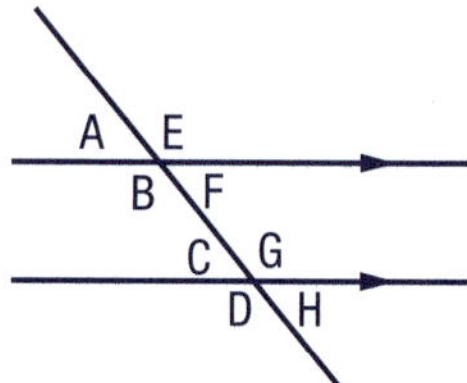

c

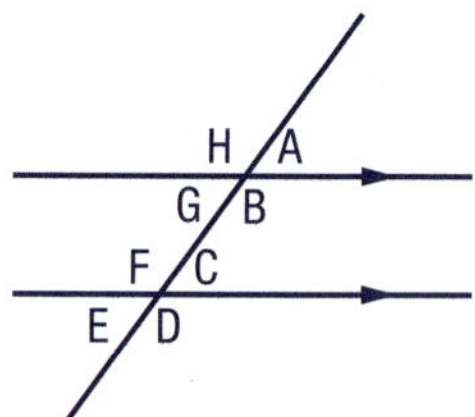

d

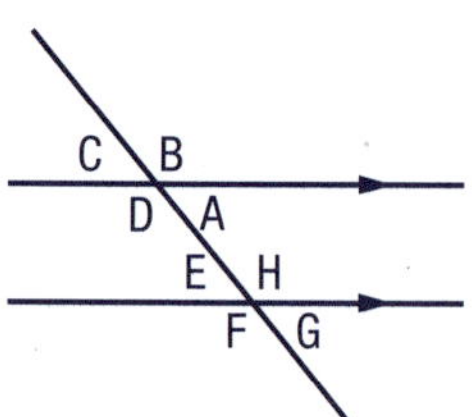

e

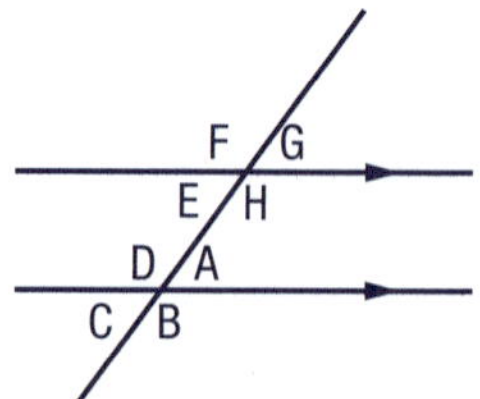

f

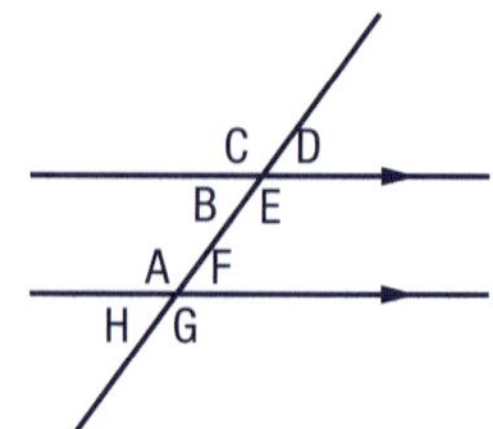

2 Which angle is alternate to *Y*?

a
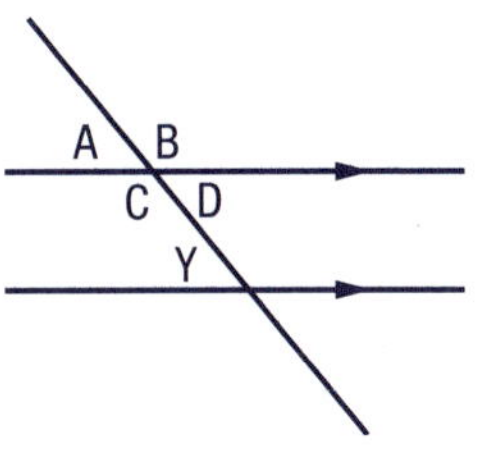

b
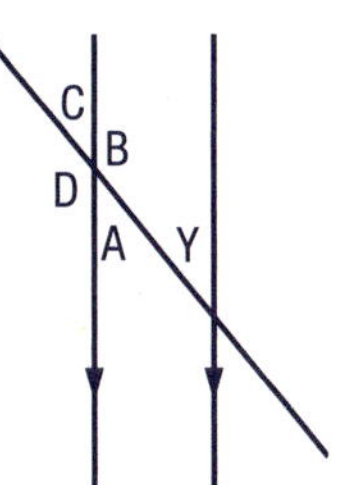

c
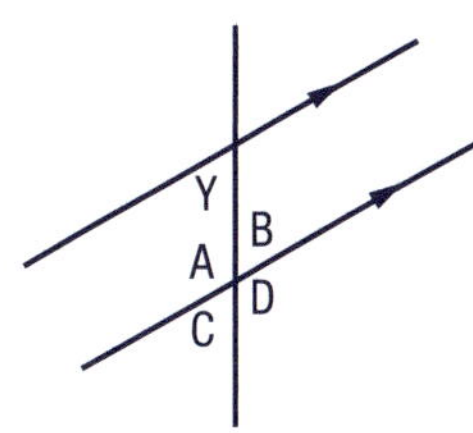

3 Which angle is co-interior to *R*?

a
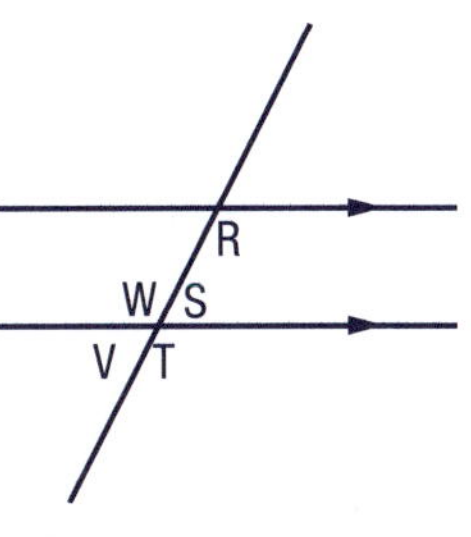

b
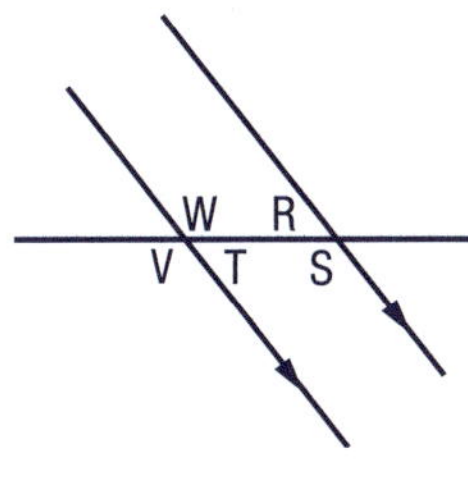

c
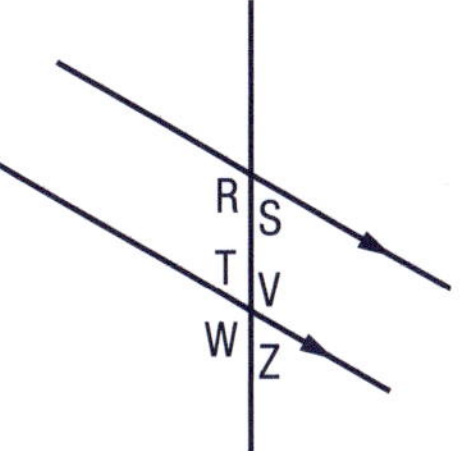

4 In your workbook, sketch a pair of parallel lines crossed by a transversal.

a Use dots to mark the location of two alternate angles.

b Use crosses to mark the location of two corresponding angles.

c Use squares to mark the location of two co-interior angles.

5 Calculate the value of the pronumerals in each of the following examples. Explain how you worked out your answer in each case.

a
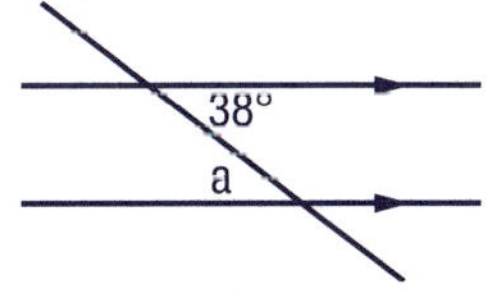

b
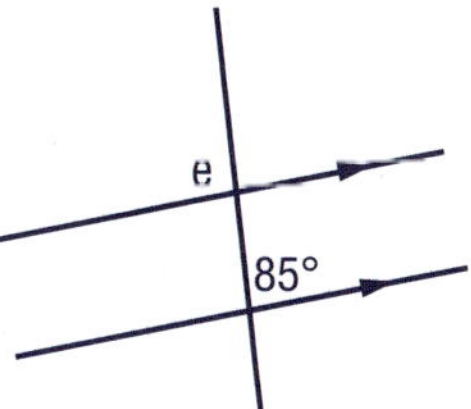

c
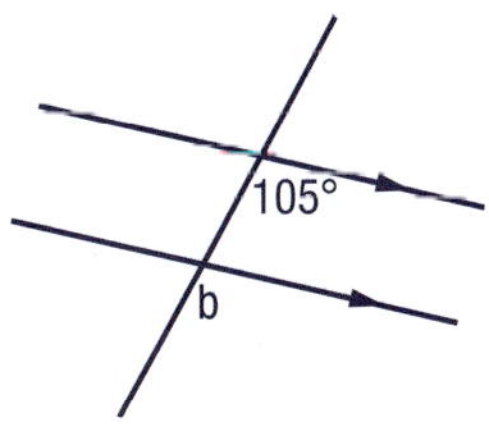

d
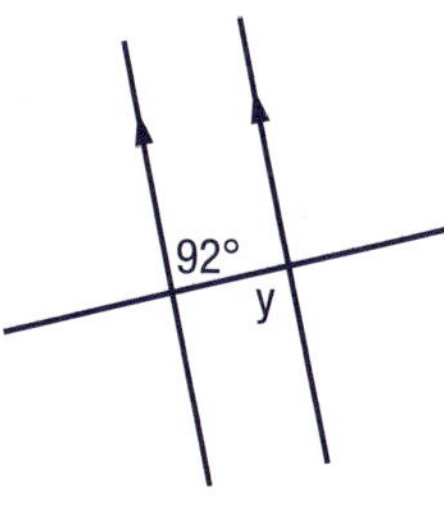

e
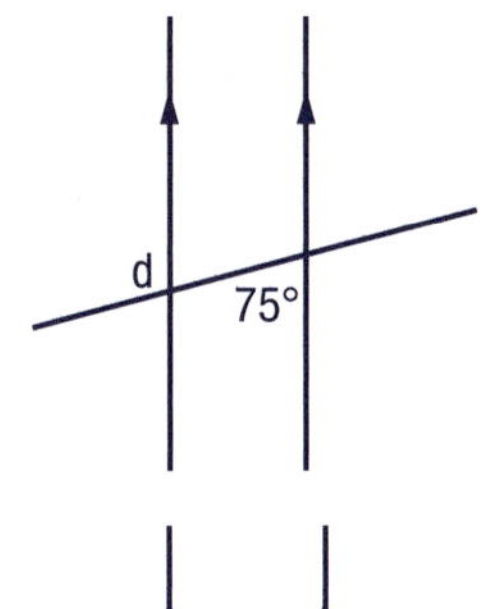

f
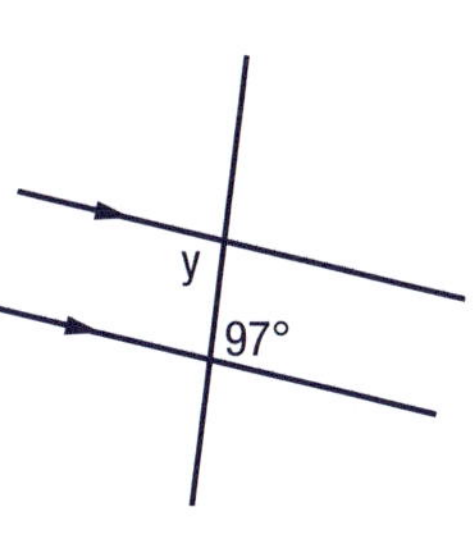

g
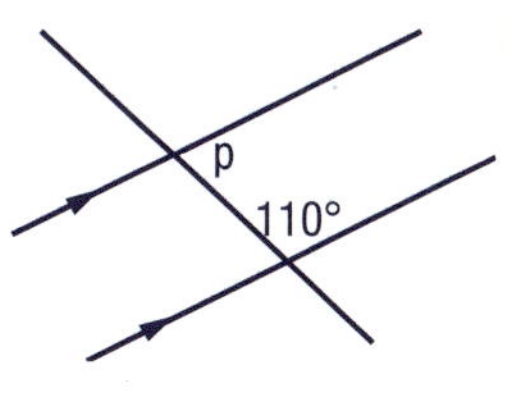

h
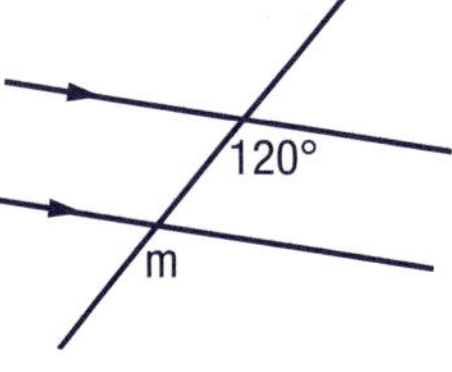

i
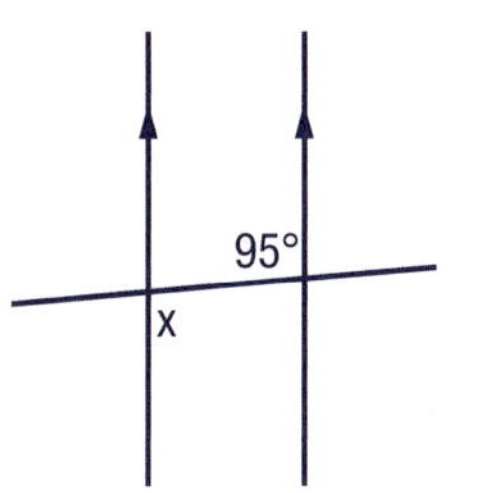

j
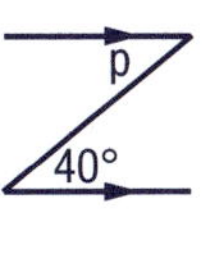

k
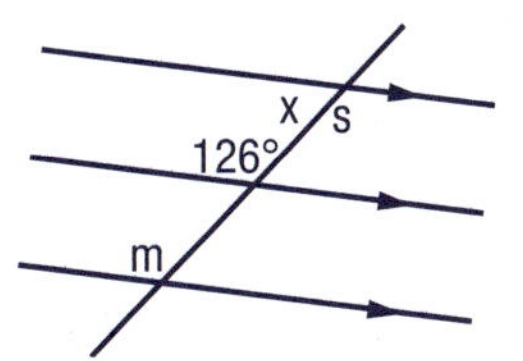

l
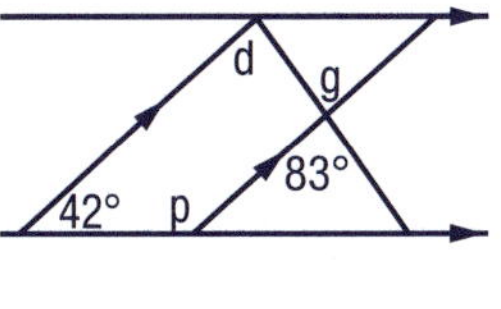

Challenge

Draw four parallel lines and cut them with a transversal. Measure and label the minimum number of angles that would enable a friend to calculate all the missing angles in your diagram.

Lesson 6 Polygons and angles

In Lesson 5 you learned how to calculate the size of the angles formed when parallel lines are crossed by a transversal. We can also calculate the sum of the internal angles of a polygon of any shape without using a protractor.

1 Remember that the internal angles of a triangle sum to 180°. Copy the following table into your workbook and complete it.

Polygon	Number of sides	Number of triangles within shape	Angle sum
	3	1	$1 \times 180° = 180°$
	4	2	$2 \times 180° = 360°$

Help Box

The number of triangles (all connecting at a vertex) in a polygon equals the number of sides in the polygon minus 2.

The number of triangles in a polygon $\times$ 180° = the sum of the interior angles of the polygon.

So $(n - 2) \times 180°$ = the sum of the interior angles of a polygon when n equals the number of sides on the polygon.

2 Draw the following polygons into your workbook and use the formula to calculate the sum of the internal angles of each shape.

a regular pentagon

b regular hexagon

c irregular quadrilateral

d irregular hexagon

e regular octagon

f irregular octagon

3 Calculate the sum of the internal angles of:

a a septagon   b a nonagon   c a decagon

4 Use the formula to find the value of the unknown angles in each of the following polygons.

a
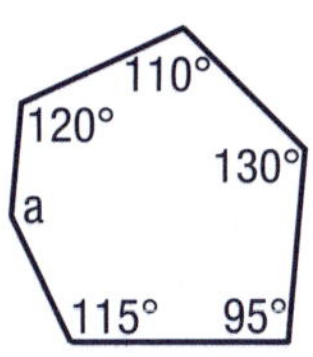

b
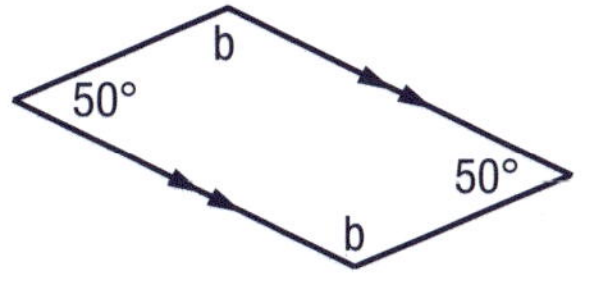

c
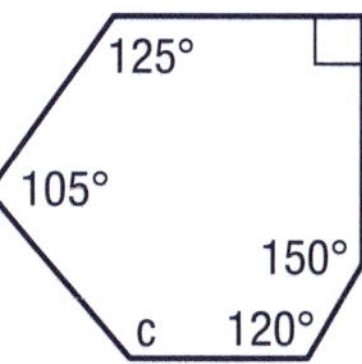

d
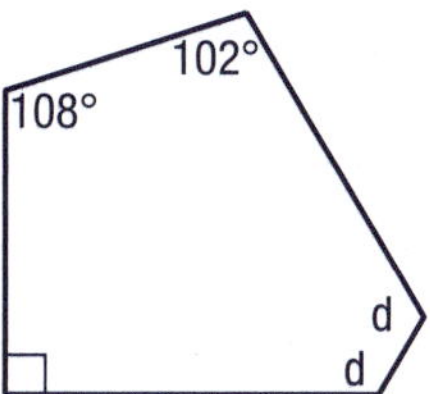

e
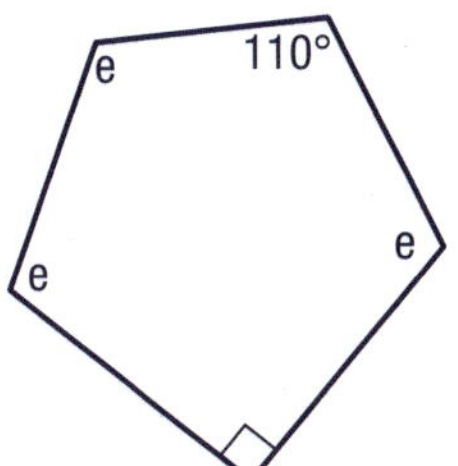

f
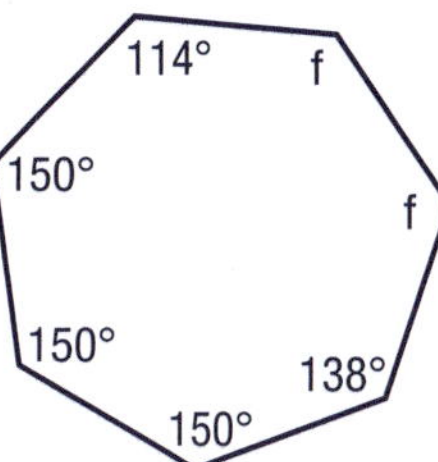

g
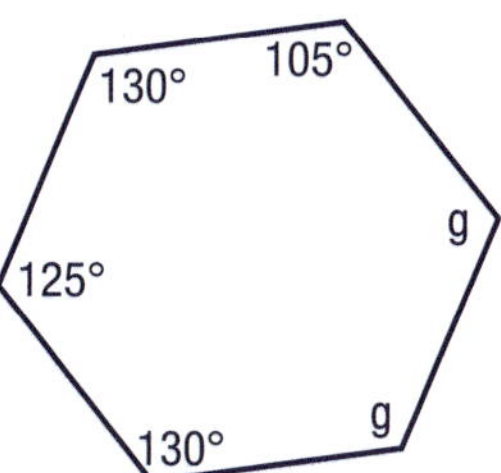

h
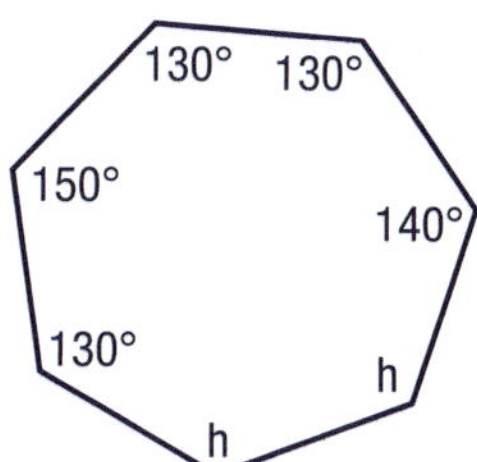

Help Box

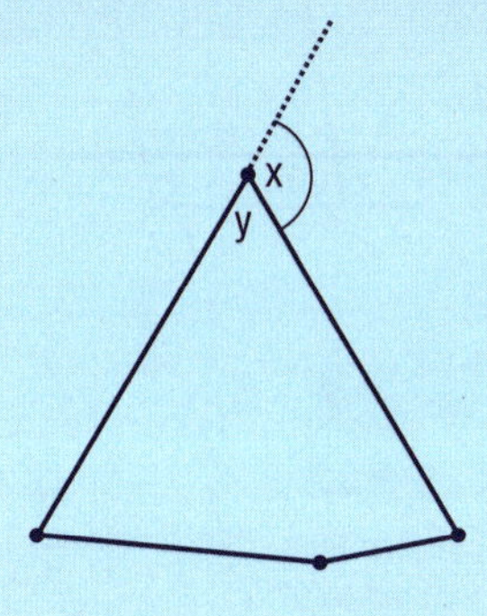

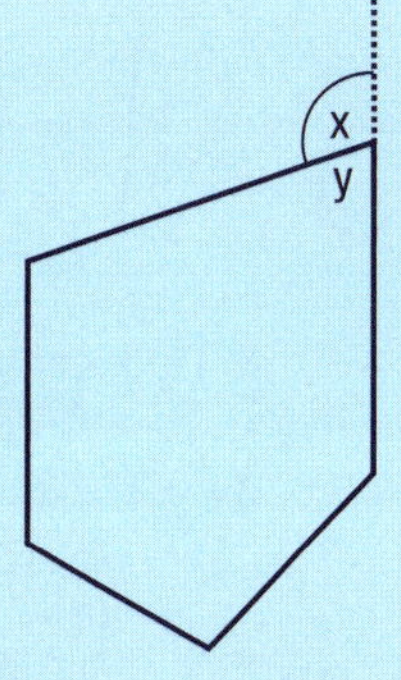

The angles outside a polygon are called the **external angles**. The external and interior angles at the same vertex total 180°. They are supplementary angles.

Angles labelled x are external angles.

Angles x and y total 180° and are supplementary angles.

5 Use your knowledge of interior and exterior angles to find the missing angles in each of the following diagrams.

a

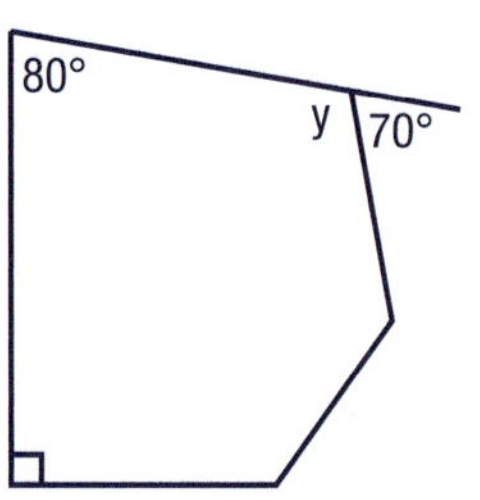

b

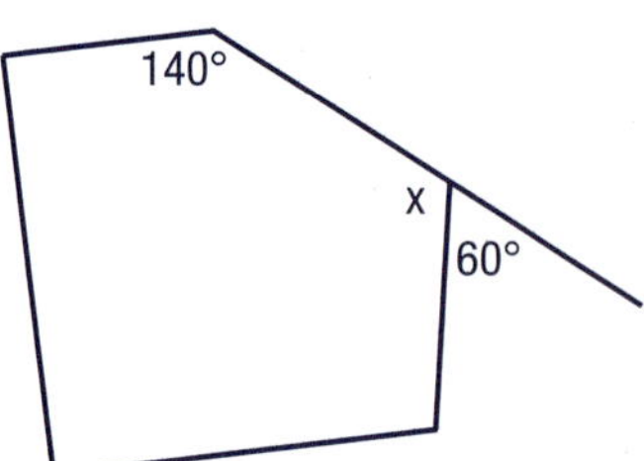

c

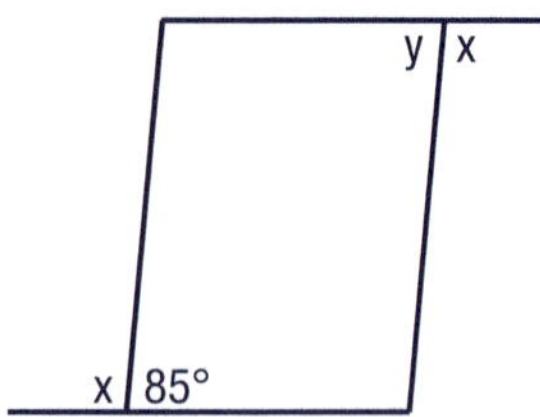

d

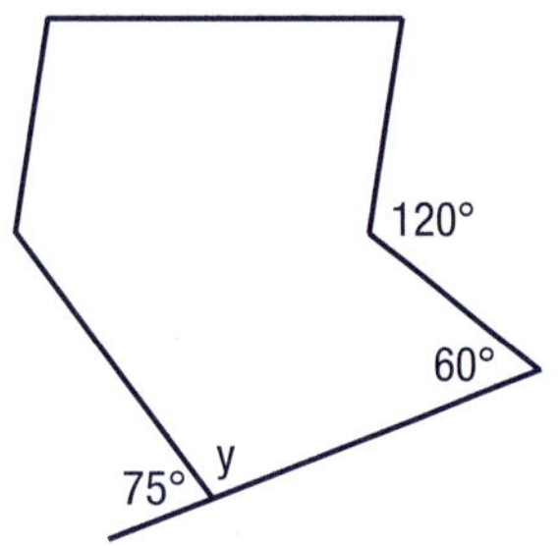

e

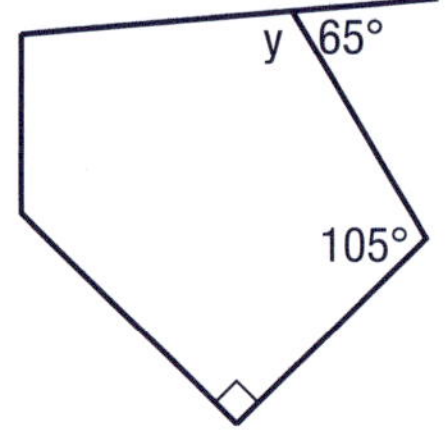

f

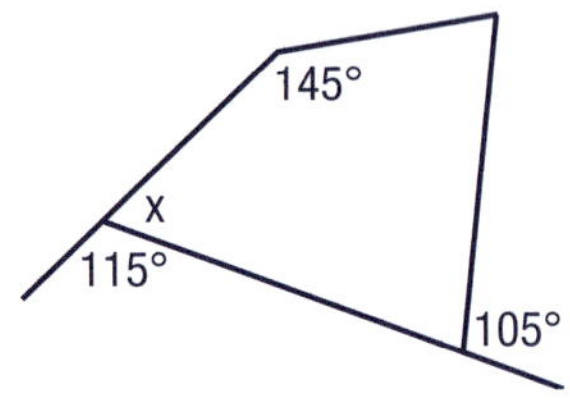

6 Copy the following polygons into your workbook and use your knowledge of polygons to find the missing angles in each.

a

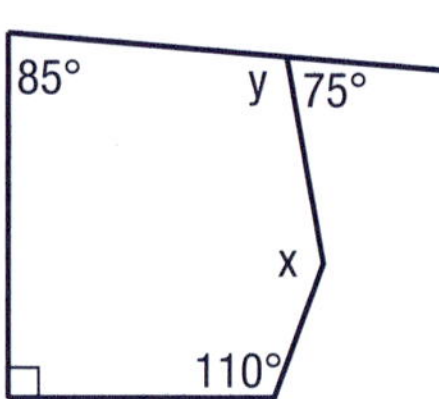

b

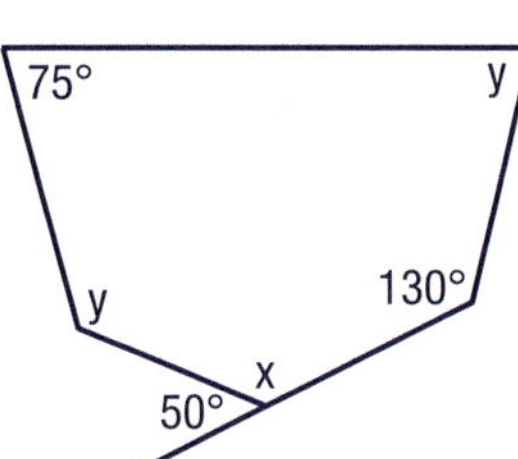

c

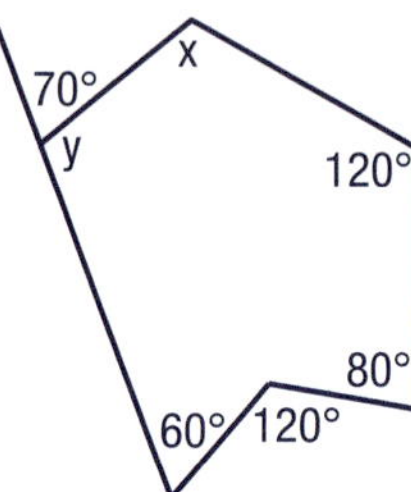

d

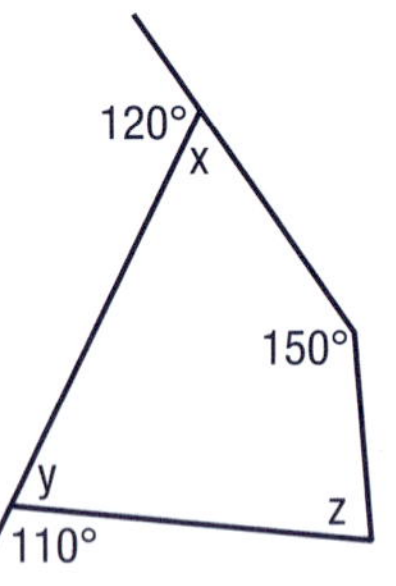

e

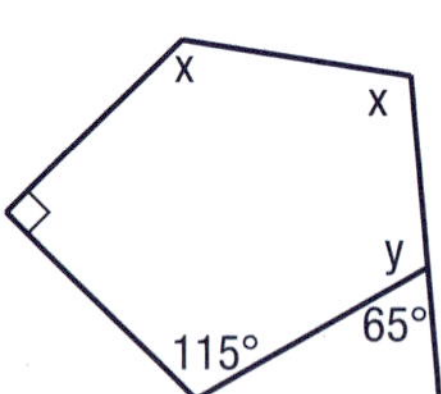

f

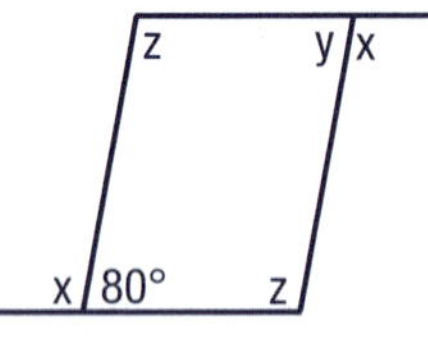

g

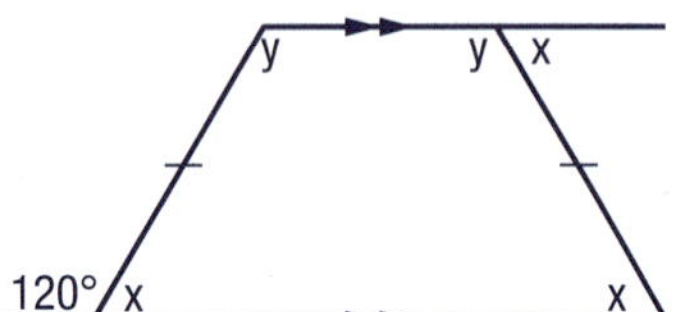

Challenge

Investigate whether the formula $(n - 2) \times 180°$ can be used to also find the sum of the internal angles of concave polygons.

This is an example of a concave pentagon.

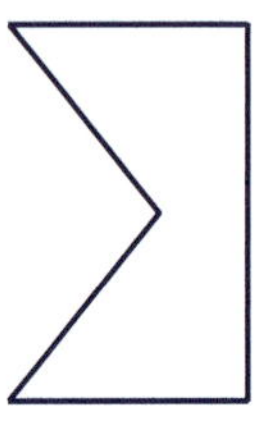

Lesson 7 Fractions and percentages

In Year 7 you learned about fractions and percentages. A percentage is a special fraction that always has a denominator of 100. Fractions are commonly used to calibrate certain tools such as spanners and nuts and bolts, and to size and measure other building materials including timber.

Help Box

The mix of imperial and metric measurement systems used internationally means that some tools are labelled in fractions of an inch (imperial system) and others are labelled in millimetres (metric system).

One inch is approximately equal to 2.5 centimetres.

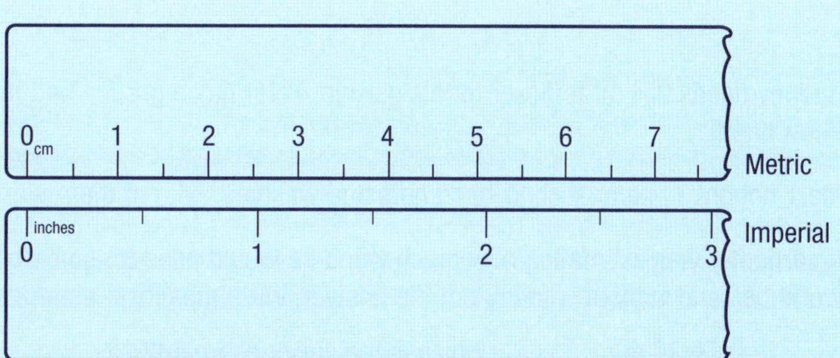

Revise your knowledge of fractions by answering the following questions.

1 Which spanner is the larger in each pair?

a $\frac{5}{16}$ inch or $\frac{1}{2}$ inch
b $\frac{1}{4}$ inch or $\frac{5}{16}$ inch
c $\frac{11}{32}$ inch or $\frac{1}{8}$ inch
d $\frac{9}{16}$ inch or $\frac{5}{8}$ inch
e $\frac{3}{4}$ inch or $\frac{11}{16}$ inch
f $1\frac{7}{8}$ inch or $1\frac{13}{16}$ inch

2 Write all the fractions in question 1 in order from smallest to largest.

3 Samson has $\frac{2}{3}$ kg of nails and he buys another $\frac{1}{2}$ kg from the market. How many kilograms of nails does he now have?

4 Jenny wants to paint a room. She has $1\frac{1}{4}$ L of paint in one tin, $\frac{2}{3}$ L in another and about $2\frac{3}{8}$ L in a third.

a How much paint does Jenny have altogether?

b How much more paint does she need to have 5 L?

c If Jenny's father said she was not allowed to use the tin that contains $\frac{2}{3}$ L, how any litres short of 5 L will she be?

5 A stallholder is selling bundles of eight pieces of wood each $2\frac{1}{3}$ metres long. Otto needs six lengths of wood each $1\frac{3}{4}$ metres long to build some shelves.

a What is the total length of wood in one bundle?

b What is the total length of the wood that Otto needs?

c How much wood will Otto have left per piece if he cuts a $1\frac{3}{4}$ metre length from a piece of wood $2\frac{1}{3}$ metres long?

d How much wood will he have left over altogether if he only cuts six of the eight pieces?

Challenge

If the stallholder sells individual lengths of wood $2\frac{1}{3}$ metres long for K2.20 or bundles of six for K11, which would be the cheapest way for Otto to buy the wood he needs? Show your working out.

6 Harold has a piece of wood $3\frac{3}{8}$ metres in length. It takes $\frac{2}{3}$ of a metre of wood to build a toy car.

a How many toy cars can Harold build from the length of wood?

b What length of wood will be left?

7 Antonia sized some bolts himself and put them into 12 bags labelled with the fraction of an inch that he considers they are. Without realising, he has used equivalent fractions to label the bags, meaning that some measurements have been duplicated. Determine which bags should be combined.

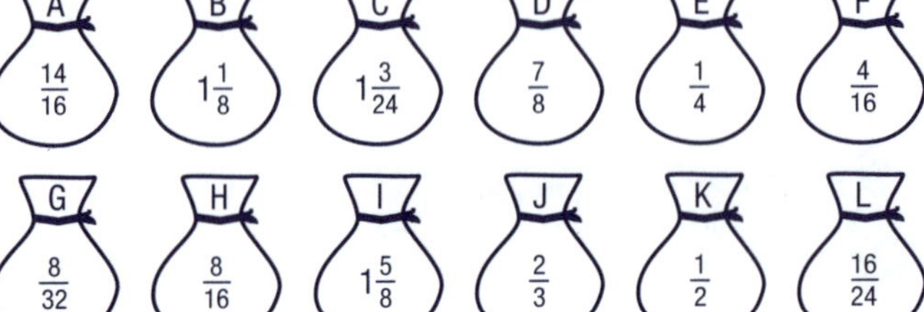

8 Rodney needs $2\frac{2}{3}$ L of fertiliser for his garden. Peter has a garden bed that is $1\frac{3}{4}$ the size of Rodney's. How much fertiliser will Peter need?

9 Mary bought $1\frac{1}{4}$ kg of second-hand nuts and washers. $\frac{4}{5}$ kg of them were usable. How many kg were not usable?

10 A carpenter was estimating how much wood he would need for some building projects. Copy the fractions into your workbook and without working out the answer, write the closer estimate for each calculation.

a	$1\frac{5}{8}$ m + $1\frac{2}{3}$ m	which is the closer estimate?	$3\frac{1}{3}$ or $3\frac{3}{4}$ metres
b	$3\frac{5}{16}$ m + $\frac{7}{8}$ m	which is the closer estimate?	$4\frac{1}{2}$ or 4 metres
c	$4\frac{7}{10}$ m − $\frac{3}{8}$ m	which is the closer estimate?	$4\frac{1}{8}$ or $4\frac{3}{4}$ metres
d	$2\frac{3}{10}$ m − $\frac{1}{3}$ m	which is the closer estimate?	2 or $1\frac{7}{10}$ metres

11 Calculate the accurate answers to question 10 to check your estimations.

12 In a toolbox there are 16 different tools. Six of the tools are new. One-third of the new tools and half of the old tools are woodworking tools.

a How many woodworking tools are in the toolbox?

b What fraction of the tools in the toolbox are old?

c What fraction of tools in the toolbox are not woodworking tools?

d Draw a diagram to describe your reasoning for these three questions.

Revise your knowledge of percentage by answering the following questions.

13 Convert the following fractions to percentages.

a	$\frac{39}{100}$	b	$\frac{9}{50}$	c	$\frac{7}{20}$	d	$\frac{3}{10}$	e	$\frac{3}{4}$
f	$\frac{2}{5}$	g	$\frac{48}{150}$	h	$\frac{7}{10}$	i	$1\frac{1}{4}$	j	$\frac{1}{3}$

14 The store sold 1200 L of paint in one week. Work out how many litres of each colour was sold if:

a	6% was black	b	8% was green	c	10% was brown
d	16% was yellow	e	25% was blue	f	35% was white

15 A hardware store is having a sale and taking 20% off all its stock. What will be the sale price of each item shown?

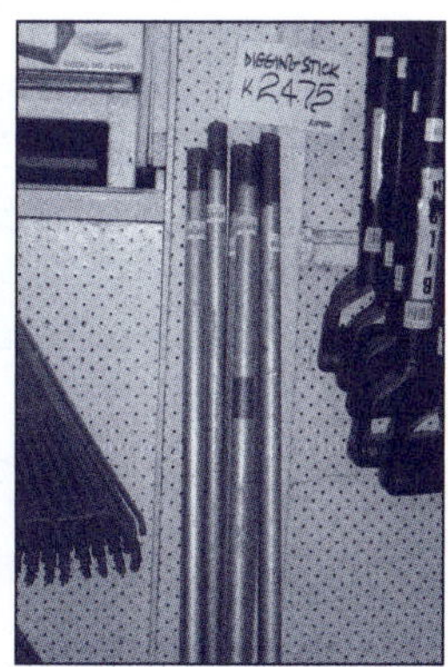

a	knives @ K5.38	**b**	wheelbarrow @ K162.62	**c**	shovels @ K15.60
d	builder's shovels @ K15.40	**e**	drop saw @ K1194.04	**f**	digging stick @ K24.75
g	handsaw @ K17.15	**h**	hacksaw @ K13.75		

Lesson 8 Comparing common fractions and decimals

A shifting spanner is a tool used to loosen or tighten nuts or bolts.

The 'jaw' (the part into which the nut or bolt goes) can be adjusted in size to fit different sized nuts and bolts. Spanners are labelled according to the size of the bolt that they fit. Different international number systems have resulted in some spanner sizes being labelled as a fraction of 1 inch while others are in millimetres.

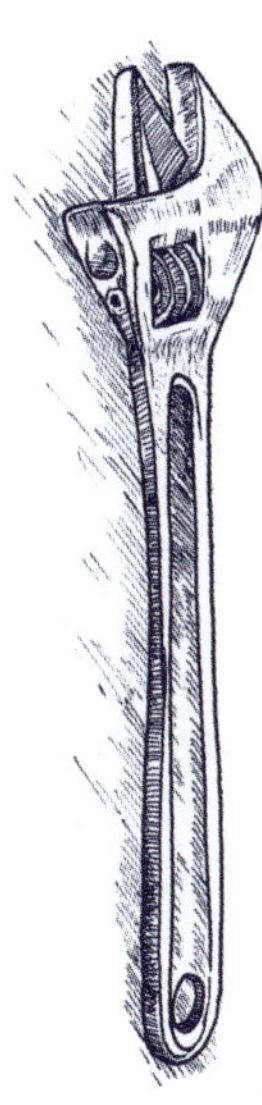

Help Box

To change a decimal into a common fraction, place the decimal number above the appropriate denominator and then simplify the fraction.

For example, $0.250 = \frac{250}{1000}$ (to simplify divide numerator and denominator by 250)

$= \frac{1}{4}$

You may not always see the quickest way to simplify the fraction. This means it might take extra steps to convert the fraction to its simplest form.

For example, $0.250 = \frac{250}{1000}$ (to simplify divide numerator and denominator by 10)

$= \frac{25}{100}$ (to simplify divide numerator and denominator by 25)

$= \frac{1}{4}$

1 Copy this table into your workbook and convert the decimals to common fractions in their simplest form.

Spanner sizes given as a fraction of an inch Decimal to fraction conversion chart								
Decimal fraction	0.25	0.313	0.375	0.445	0.5	0.625	0.75	0.875
Common fraction	$\frac{1}{4}$							

2 Write these decimals as common fractions in their simplest form.

a 0.6 b 0.2 c 0.85 d 0.45 e 0.05 f 0.12
g 0.18 h 0.46 i 0.37 j 0.93 k 0.36 l 0.24

3 Copy this table into your workbook and convert each decimal number to a mixed number with a fraction in its simplest form.

Spanner sizes given as a fraction of an inch Decimal to fraction conversion chart						
Decimal fraction	1.125	1.25	1.625	1.6875	1.875	2.050
Common fraction						

4 Write these decimals as mixed numbers with the fraction in its simplest form.

a 1.8 b 2.4 c 6.08 d 3.75 e 8.28 f 14.36
g 4.085 h 2.146 i 5.035 j 6.125 k 23.028 l 10.264

The three different international measurement standards for spanners are known as:

- British Standard Fine (BSF)
- British Standard Whitworth (Wworth)
- Across Flat (AF)

5 Spanner sizes relate to a fraction of one inch. Convert all measurements in each of the following sets to either decimals or common fractions, then determine which spanner will handle the largest bolt.

a $\frac{3}{16}$ Wworth $\frac{1}{4}$ AF 0.394 inch setting
b 0.6 inch $\frac{5}{16}$ Wworth $\frac{3}{8}$ BSF
c 0.63 inch $\frac{3}{4}$ AF $\frac{11}{16}$ Wworth
d $\frac{13}{16}$ AF 0.748 inch $\frac{3}{8}$ BSF
e 0.820 inches $\frac{7}{16}$ Wworth $\frac{1}{2}$ BSF

Challenge

Write three different mixed numbers with fractions that have different denominators and have values between 2.5 and 2.6.

Help Box

To change a common fraction into a decimal, divide the numerator by the denominator.

$\frac{3}{8}$ means $3 \div 8$

$$\begin{array}{r} 0.375 \\ 8\overline{)3.0} \\ 2\,4 \\ \hline 60 \\ 56 \\ \hline 40 \\ 40 \\ \hline 0 \end{array}$$

$\frac{3}{8} = 0.375$

6 Convert each of the following to a decimal fraction (to three decimal places).

a $\frac{1}{2}$ b $\frac{3}{4}$ c $\frac{1}{5}$ d $\frac{5}{8}$ e $\frac{7}{10}$ f $\frac{3}{8}$

g $\frac{2}{9}$ h $\frac{4}{5}$ i $\frac{9}{20}$ j $\frac{2}{3}$ k $\frac{4}{15}$ l $\frac{6}{11}$

7 Change each set of fractions to decimals and then write the fractions in each set from smallest to largest.

a $\frac{1}{2}, \frac{3}{10}, \frac{4}{5}$ b $\frac{3}{5}, \frac{4}{9}, \frac{4}{7}$

c $\frac{2}{7}, \frac{5}{11}, \frac{3}{8}$ d $\frac{3}{10}, \frac{4}{11}, \frac{5}{13}$

e $\frac{4}{5}, \frac{7}{8}, \frac{2}{3}$ f $\frac{3}{7}, \frac{2}{5}, \frac{1}{4}$

g $\frac{6}{11}, \frac{4}{9}, \frac{3}{7}$ h $\frac{5}{8}, \frac{7}{13}, \frac{25}{32}$

Challenge

A decimal that repeats is called a **recurring decimal**.

The fraction $\frac{1}{3}$ makes the recurring decimal 0.333333 OR $0.\dot{3}$.

We usually write recurring decimals by placing a dot above the repeating number(s).

Which denominators between 1 and 10 convert to recurring decimals when the numerator is 1?

8 Write a common fraction that occurs between the pairs of numbers.

a 0.25 and 0.78 b 0.13 and 0.45 c 0.65 and 0.9 d 0.03 and 0.1

e 0.13 and 0.2 f 0.23 and 0.31 g 2.02 and 2.1 h 3.8 and 3.08

9 Write a decimal fraction that occurs between the pairs of numbers.

a $\frac{1}{8}$ and $\frac{3}{8}$ b $\frac{1}{2}$ and $\frac{3}{4}$ c $\frac{1}{5}$ and $\frac{7}{10}$ d $\frac{1}{4}$ and $\frac{3}{5}$

e $\frac{1}{3}$ and $\frac{7}{8}$ f $\frac{4}{10}$ and $\frac{5}{8}$ g $\frac{4}{5}$ and $\frac{7}{8}$ h $\frac{3}{10}$ and $\frac{3}{8}$

Lesson 9 Ratio

Some tools rely on gears (or cogs) to increase or decrease the rotational speed in a set of wheels. In this example there are 54 teeth on the front gear and 27 teeth on the rear gear. This means that every time the larger gear goes around once, the smaller gear goes around twice. We can use ratio to compare quantities. The gear ratio for this set of wheels is 54:27.

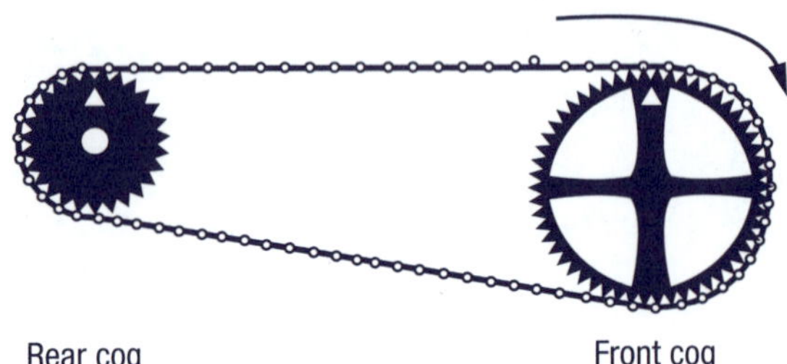

1 Write a ratio for each of the following.

a the ratio of teeth on gears

b the ratio of hammers to spanners

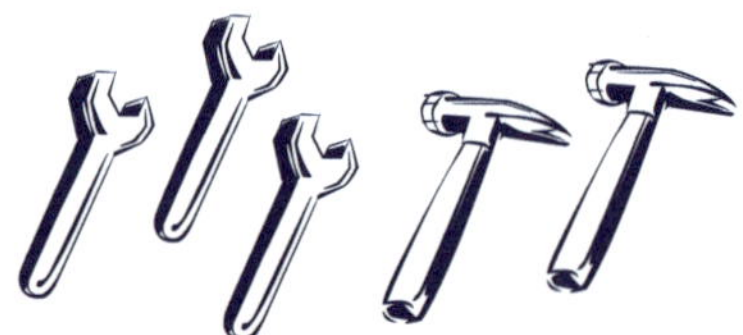

c the ratio of nuts to bolts

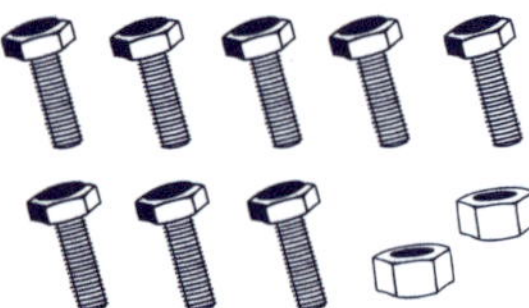

d the ratio of the triangle's height to its base length

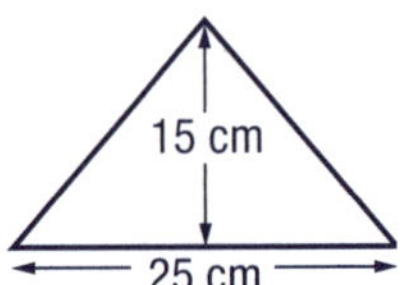

e the ratio of squares to triangles

f the ratio of fertiliser to water

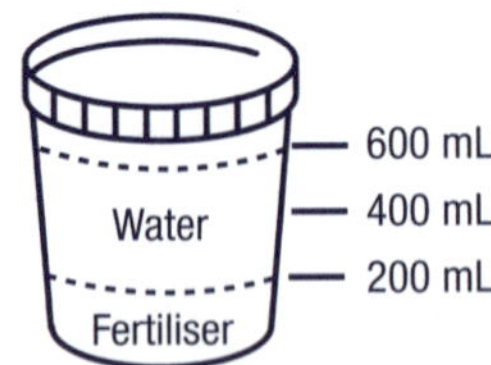

2 A particular colour is made by mixing 4 litres of yellow paint with 7 litres of orange paint. Find the ratio of:

a yellow paint to orange paint

b yellow paint to the total amount of paint in the mixture

c orange paint to yellow paint

d orange paint to the total amount of paint in the mixture

3 Every panel in a storage cupboard requires 10 nails and 4 screws. Find the ratio of:

a nails to screws per panel

b screws to nails per panel

c screws to the total number of screws and nails in each panel

d nails to the total number of screws and nails in each panel

4 Jenny mixes 3 buckets of manure with 10 buckets of soil to make some potting mix. What ratio of manure to soil did she use?

5 Andrew built a stool and a small table. The stool used 2.5 metres of wood; the table used 4 metres of wood. Find the ratio of wood used for the stool compared to the table.

6 Tacky glue is made by mixing 3 teaspoons of Tube A with $\frac{1}{2}$ a teaspoon of Tube B. What is the ratio of Tube A contents to Tube B contents in a mixture of tacky glue?

The gear ratio 54:27 shown at the start of this lesson could also be written more simply as the ratio of 2:1 because two revolutions of the rear gear mean one revolution of the front gear.

Help Box

Ratios can be converted to their simplest form by multiplying or dividing each number by the same value. For example,

25:55 divide both numbers by 5 ⟶ 5:11
The ratio 25:55 is the same ratio as 5:11.

$4\frac{1}{2}$:7 multiply both numbers by 2 ⟶ 9:14
The ratio $4\frac{1}{2}$:7 is the same ratio as 9:14.

7 Simplify each of the following ratios.

a	2:8	b	3:15	c	4:20	d	35:10	e	12:36	f	60:24
g	21:42	h	56:80	i	77:33	j	38:114	k	6:300	l	5:105

8 Write each of the following ratios in whole numbers in their simplest form.

a $2\frac{1}{2}$:7 b $1\frac{1}{4}$:5 c $6\frac{1}{3}$:7 d $1\frac{3}{4}$:4 e $4\frac{1}{4}$:5 f $10\frac{1}{4}$:3

9 Convert each of the following ratios to their simplest whole number form by deciding what the common fraction and/or mixed number needs to be multiplied by to make a whole number. Question a has been done as an example.

a $\frac{3}{5}$:15 $\frac{3}{5} \times \frac{5}{3} = 1$ Therefore $\frac{3}{5}$:15 is the same ratio as $(\frac{3}{5} \times \frac{5}{3})$:$(15 \times \frac{5}{3})$ or 1: 25.

b $\frac{3}{5}$:10 c $\frac{2}{3}$:5 d $\frac{2}{3}$:6 e $\frac{3}{8}$:6 f $2\frac{5}{8}$:21 g $4\frac{2}{3}$:2

Help Box

A ratio can also be thought of as a fraction.

For example, if gear X spins twice while gear Y spins four times, the ratio is 2:4 which can be simplified to 1:2.

The ratio 2:4 can also be written as the fraction $\frac{2}{4}$ and then simplified to $\frac{1}{2}$.

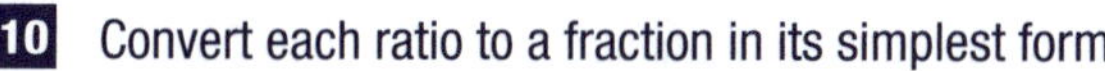

10 Convert each ratio to a fraction in its simplest form.

a	3:9	b	5:25	c	12:66
d	30:21	e	12:108	f	7:7
g	24:60	h	49:84	i	18:56
j	38:76	k	12:94	l	6:75

11 Simplify the ratios.

a	5:100	b	15:75	c	84:6
d	$3\frac{1}{2}$:14	e	$\frac{4}{10}$: 6	f	$\frac{2}{5}$: 8
g	$2\frac{1}{2}$:10	h	$4\frac{3}{8}$:6	i	1.5:3
j	3:0.75	k	2.5:3	l	2.2:3.5

Lesson 10 Problems in the tool shed

Help Box

If the ratio and one quantity in a comparison are known, the other quantities can be found.

For example, the ratio of water to chemical concentrate in a garden spray is 5:3.

To calculate how many litres of chemical concentrate are needed with 20 L of water, mentally multiply the water side of the ratio (5) until 20 is obtained ($\times$ 4), then multiply the chemical concentrate by the same amount (3 $\times$ 4).

So 12 L of concentrate are needed for every 20 L of water.

Show your working out for each of the following ratio problems.

1 A particular shade of paint is made by mixing blue and red paint in the ratio 4:7. If Sam begins with 800 mL of blue paint, how much red must he add?

2 Rosa mixed the meat and cereal for the animal feed in the ratio 3:11. If she used 180 g of meat, how much cereal did she use?

3 A tile glue is made by combining the contents of Tube A with the contents of Tube B in a ratio of 2:5. If 6 mL of Tube A is used, how much of Tube B is needed?

4 A plaster contains a mix of powder and water in a ratio of 5:4. If 1.8 L of plaster has been mixed, what quantity of each ingredient has been used?

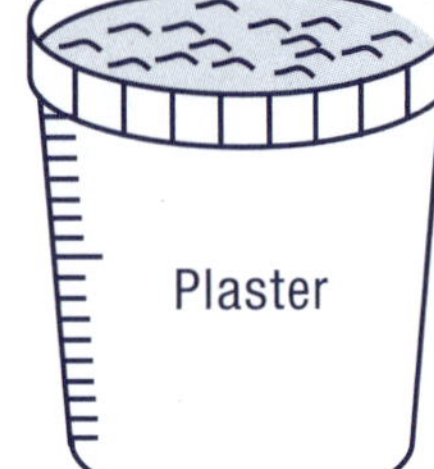

Powder to water 5:4 ratio

5 A concreting mixture uses the ratio 3:2:1 meaning 3 parts screenings, 2 parts sand and 1 part cement. If I wanted to make 18 shovelfuls of concrete using this ratio, how many shovelfuls of each ingredient would I need?

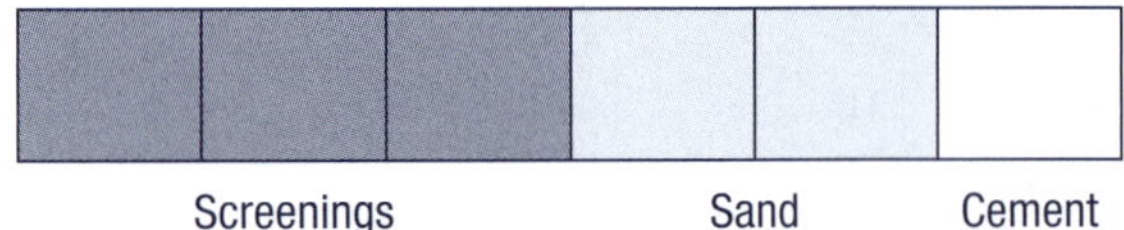

6 A certain garden mulch comprises bark, leaves and sawdust in the ratio 5:2:1. If Otto wants to make 16 kg of mulch, how many kilograms of each item will he need?

Challenge

Use a mulch recipe of bark, leaves and sawdust in a ratio of 7:3:2 to calculate the quantity of bark and leaves required if Tracy began with 3 kg of sawdust.

Solve the following problems and write your answer in a manner appropriate to each situation.

7 What fraction of a kilogram is 50% of $\frac{3}{4}$ kg of nails?

8 What percentage remains of a piece of wood 2.7 m long if I have used $\frac{2}{3}$?

9 If the ratio of chemicals to water in a mixture is 2:4, what fraction of the mixture is water?

10 If 0.8 of the tools in the toolbox are old, what percentage is new?

11 What quantity of paint remains if $\frac{5}{8}$ has been used from a tin containing 4 L?

12 Copy the table into your workbook and complete it.

Percentage	Fraction	Decimal	Ratio
60%			
	$\frac{1}{2}$		
		0.25	
			3:4
15%			
	$\frac{1}{3}$		

Help Box

Decimals, fractions, percentages and ratios are related to each other.

For example, 0.6 can be expressed as $\frac{6}{10}$, 60% and 6:10 as they are all equal.

They can be used interchangeably to suit the situation.

Challenge

Fegsley and Mary each built a small model of a belltower. The ratio of the height of Fegsley's tower to Mary's tower is 5:6. If Fegsley's tower is 34 cm tall, how tall is Mary's tower?

Learning Unit  Additional Learning, Revision and Assessment

Strand: Number and Application

Directed Numbers	Outcome 8.1.7	Apply directed numbers in problem solving

Lesson 1: Additional learning

Focus: Multiplying and dividing directed numbers

Lesson 2: Revision

Fractions, decimals, percentages and ratios
Simple algebraic expressions
Angles of polygons
Directed numbers

Lesson 3: Assessment Task 1

Practical investigation—group work
Assessment Task 1 will assess learning outcomes 8.1.1, 8.1.3 and 8.1.5 and 8.4.7.

40 marks

Lesson 4: Assessment Task 2

Test of basic skills and routine applications
Assessment Task 2 will cover the learning outcomes from across the topic 'Gadgets'.
The test will assess the extent to which students can:

- Demonstrate an understanding of the mathematical concepts
- Correctly choose and apply mathematical techniques to solve problems

60 marks
Total 100 marks

Lesson 1 Additional learning

Multiplying and dividing directed numbers

In Unit 1 'Mobile Phones' you learned some simple rules to use when adding and subtracting directed numbers. This lesson will investigate the rules that apply to multiplying and dividing directed numbers.

1 Copy the following table into your workbook and find a pattern that helps you to fill in the gaps.

×	3	2	1	0	–1	–2	–3
3							
2							
1							
0							
–1							
–2							
–3							

2 Copy and complete these sentences by using the information in your table.

a When multiplying two numbers with the same sign (both positive or both negative) the answer is always ____________.

b When multiplying two numbers with different signs the answer is always ____________.

3 Calculate:

a -7×4 b -5×-8

c 3×4 d 6×-5

e -10×-8 f $+7 \times -9$

g $-4 \times +6$ h $2 \times 4 \times -3$

i $+7 \times -6 \times -3$ j $5 \times -4 \times 2$

k $-9 \times -10 \times 4$ l $-8 \times 7 \times -10$

Summary of rules for + and –

(+ +) = + (– –) = +	Two like symbols make a positive
(+ –) = – (– +) = –	Two unlike symbols make a negative

Help Box

Division can be thought about as multiplying by a fraction.

Think about the equation $6 \div 2 = 3$. Another way to express this is $6 \times \frac{1}{2} = 3$

Now consider $-6 \div 2$. This can also be expressed as $-6 \times \frac{1}{2}$.

The rules for multiplication of directed numbers tell us that the answer must be negative.

So $-6 \div 2 = -3$.

Now consider $-6 \div -2$. This can also be expressed as $-6 \times -\frac{1}{2}$.

The rules for multiplication of directed numbers tell us that the answer must be positive.

So $-6 \div -2 = +3$.

4 Calculate:

a	$-14 \div +2$	b	$+15 \div -3$	c	$+32 \div +4$	d	$-16 \div -8$
e	$12 \div -4$	f	$72 \div -6$	g	$-24 \div 6$	h	$-20 \div -4$
i	$-9 \div 3$	j	$120 \div -3$	k	$22 \div 11$	l	$-1 \div -1$

5 Copy and complete these rules.

Rules for multiplying and dividing directed numbers

$+ \times + =$ ___	Two like symbols make a positive	$+ \div + =$ ___
$- \times - =$ ___		$- \div - =$ ___
$+ \times - =$ ___	Two unlike symbols make a negative	$+ \div - =$ ___
$- \times + =$ ___		$- \div + =$ ___

Help Box

The rules for the order of operations also apply to directed numbers.

The rules are:

1. brackets first
2. then multiplication and division in order of appearance from left to right
3. finally, addition and subtraction in order of appearance from left to right.

For example,	$-12 + (-2 \times 3) \div 2$
Brackets first	$-12 + (-6) \div 2$
Then $\times$ or $\div$	$-12 + -3$
Then $+$ or $-$	-15

So $-12 + (-2 \times 3) \div 2 = -15$

6 Calculate:

a $9 - 11 + 5 - 7$

b $-4 + 7 - 6 + 8$

c $-32 \div -4 + 3$

d $-2 + (-40 \div 5) \times 3$

e $(-5 + -8) \div 4 + 3$

f $44 - 8 + (-3 \times 6)$

g $6 - 5 + (11 \times -3) + -1$

h $(-7 + 19) - (-24 \div 6)$

i $-9 - (-7 \times 6)$

Challenge

Use four 4s and as many $+$, $-$, $\div$, $\times$ signs as you like to write expressions that result in negative numbers.

For example, $4 + 4 - (4 \times 4) = -8$

Lesson 2 Revision

Check your understanding of this topic using the following exercises.

Unit one

	Mary	Wesley
Peak time charges	40 toea per 50-second block	30 toea per 40-second block
Off-peak charges	15 toea per 30-second block	20 toea per 20-second block

1 Use the table of charges to calculate:

- a How many seconds can Mary talk during peak time for K4?
- b What will be the cost for Mary to talk for 6 minutes during off-peak time?
- c What is the difference in cost between a 3 minute call made by Mary during off-peak time and a 3 minute call made by Wesley during off-peak time?
- d Calculate the maximum amount of time that Wesley could talk for K5.

2 Let n be the number of people in a railway carriage. Write an algebraic expression to describe how to calculate the number of people on the train when there are:

- a 6 carriages and 3 people in the engine compartment
- b 8 carriages and only the driver in the engine compartment
- c 6 people in the engine compartment and 4 carriages

3 Simplify:

- a $8mn + 2n + 4m + 6mn$
- b $6 + 8x + 4xy - 3 + 6xy$
- c $7p - 2n - p + 6pn + 3n$
- d $10xy - 2 + 3p - 6xy + 3$

4 Simplify:

- a $5 \times 2n + 4 \times 6s$
- b $6c + 3 \times 4x - 12$
- c $k \div 5 + 5b + b$
- d $12m \div m + 3g + 6g$
- e $x \times 3 + 8n \div 3$
- f $xy - 2 \times x \div 3 \times y$

5 Expand:

- a $4(x - y)$
- b $3(3p + g)$
- c $6(5k + 2)$
- d $9(m - 3p)$
- e $x(3y - 6)$
- f $(5x + 3)y$

6 Expand and simplify:

- a $2(5 + m)$
- b $5m(3 - 6m) + 4n$
- c $8(4s - 3) + 3(s + 5)$
- d $3 + h(3 - d) + h - 4d$
- e $5 + 2(s + 4) - s + 8$
- f $2p + 5(4 - p + 8)$

7 Use the values in the table to evaluate the following expressions.

m	n	p	s	t
2	4	6	8	10

a $m + n$ b ps
c $t + 3m$ d $5m - t$
e $5(p - 2)$ f $s(2m + p)$
g $5mps$ h mst

8 Factorise:

a $3p + 18g$ b $12xy - 6xy$
c $20hj + 14hj$ d $9pr + 21r$
e $14wy + 7w$ f $15gk - 3g$
g $18pq + 2q$ h $36pr + 12r$

9 Simplify:

a $+5 - +3$ b $+4 + -3 + -5$
c $-8 - +3 + -4$ d $-11 + +5 - +4$
e $-12 + -5 + +4$ f $-10 + -3 + +8$
g $+15 + -11 + -4$ h $+32 + -3 - -18$

Unit two

10 Calculate the diameter of a circle with a radius of:

a 23 cm b 147 mm
c $\frac{3}{4}$ m d 1.95 m

11 Calculate the radius of a circle with a diameter of:

a 250 mm b 26 cm
c $4\frac{1}{4}$ m d 5.2 m

12 Calculate how far a wheel will travel in one revolution if it has a diameter of:

a 3 m b 10 cm
c 40 mm d 5 m

13 Calculate the circumference of a wheel that has:

a a diameter of 10 mm
b a radius of 25 cm
c a diameter of 50 cm
d a radius of 1 m

14 Estimate the area of the circle using the inscribed square method.

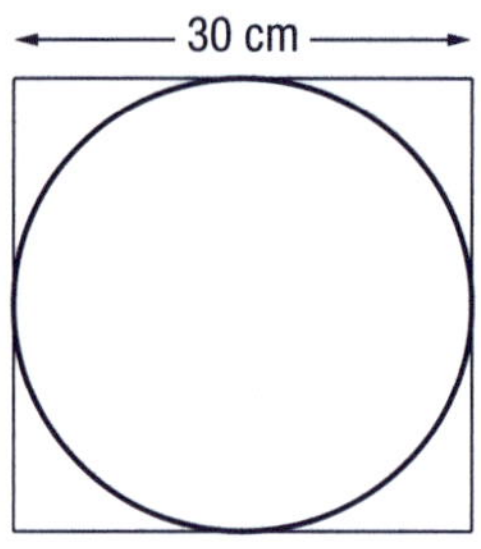

15 Calculate the area of each circle.

a
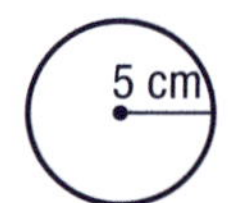

b
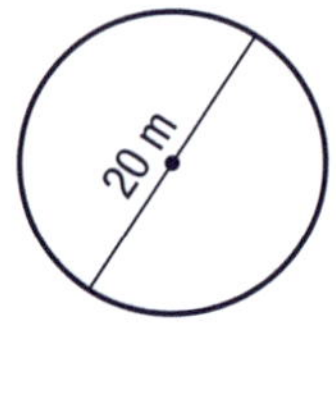

c
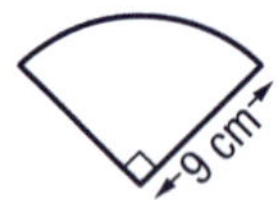

16 Find the area of these shapes.

a
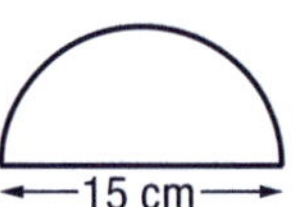

b
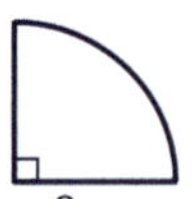

c

17 Find the area of this compound shape.

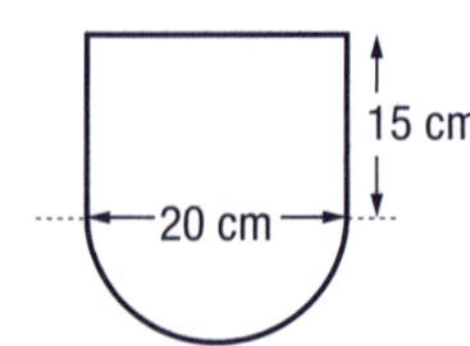

18 There are 28 students in a class. Twelve students play rugby and 14 students play tennis. There are 6 students who do not play either sport. Draw a Venn diagram to find out the number of students in the class who:

a only play rugby
b only play tennis
c play both sports
d do not play tennis

Unit three

19 Find the value of x in the following examples.

a
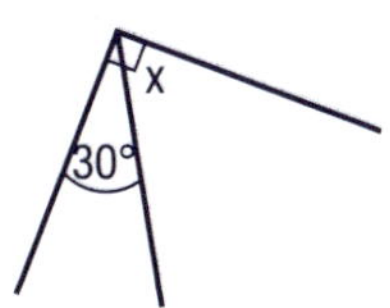

b
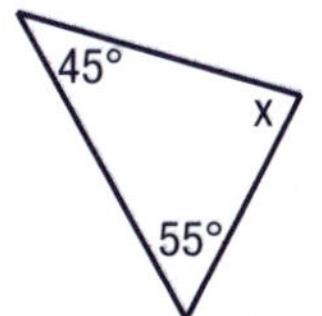

c
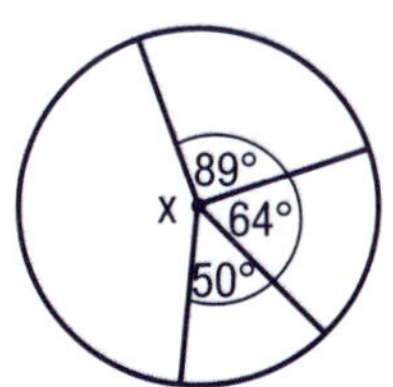

d
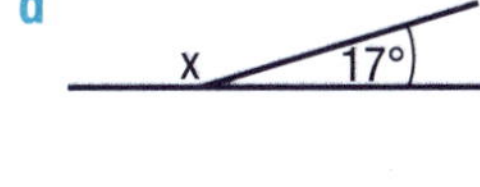

20 Use a protractor to draw three different pairs of complementary angles. Write the sizes of the angles on your diagrams.

21 What would be the supplementary angle to:

a 18°
b 45°
c 69°
d 158°
e 175°?

22 Calculate the value of the pronumerals in each of the following examples without using a protractor.

a
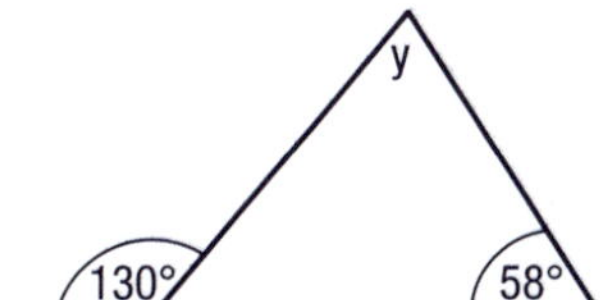

b
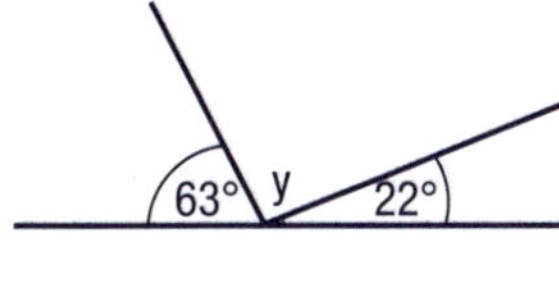

c
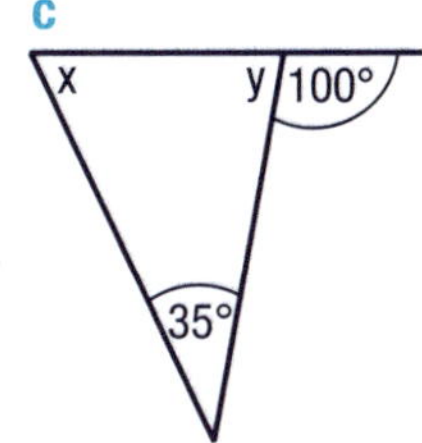

23 Calculate the value of the pronumerals in each case.

a
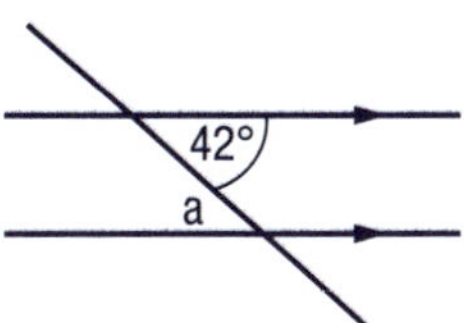

b
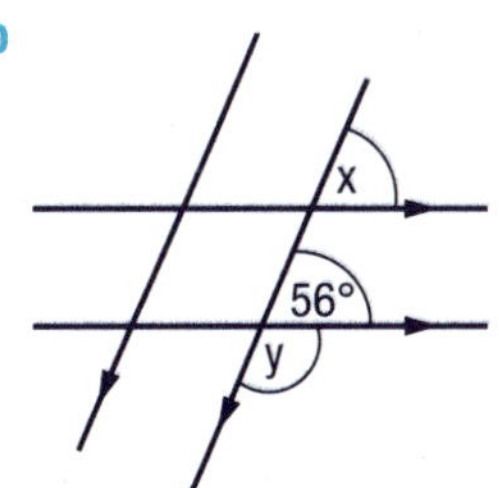

c
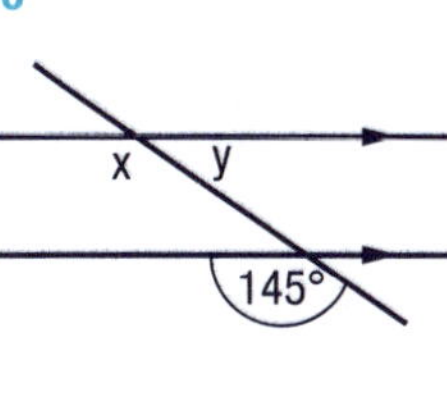

24 Draw the following polygons into your workbook and calculate the sum of the internal angles of each shape.

a irregular pentagon
b regular hexagon
c concave septagon

25 Find the value of the pronumerals in each of the following examples.

a
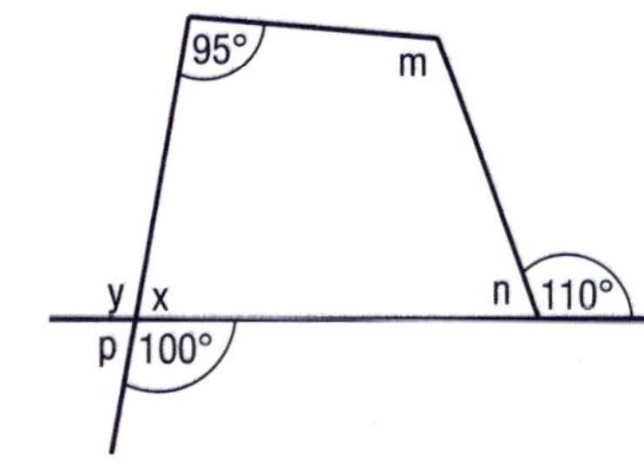

b
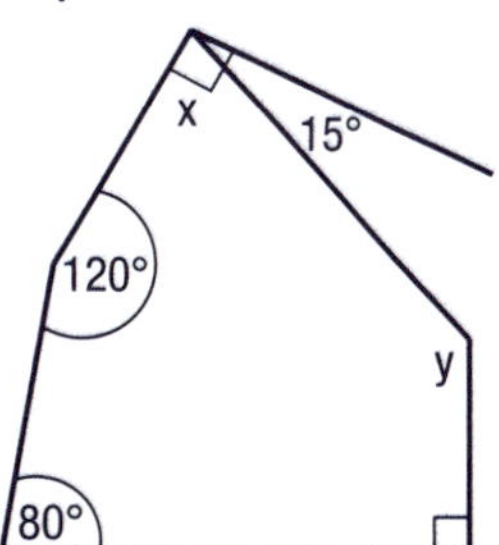

c
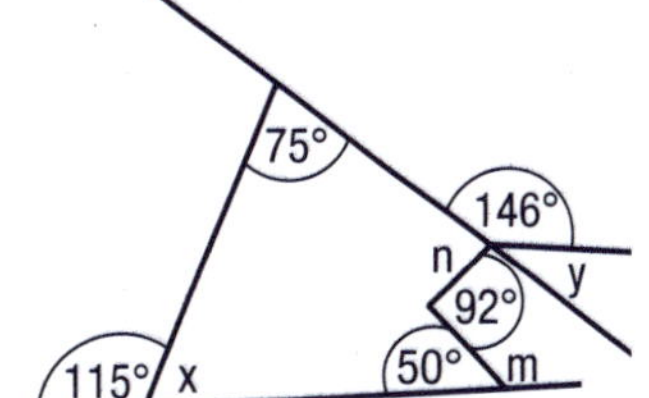

26 Solve:

a $2\frac{5}{7} + 1\frac{2}{3}$ b $4\frac{3}{4} - 2\frac{3}{5}$

c $3\frac{7}{8} + 1\frac{9}{10}$ d $3\frac{5}{8} - 1\frac{1}{3}$

27 Copy the table into your workbook and complete it.

Percentage	Decimal	Fraction	Ratio
26%			
	0.38		
		$\frac{7}{10}$	
			2:5

28 Calculate:

a 5% of 350

b $\frac{2}{3}$ of 27

c 0.75 of 60

d 15% of K3.45

e $\frac{3}{5}$ of 105

f 0.25 of 1.2 m of timber

g $\frac{9}{10}$ of 3.1 m of timber

29 Write two common fractions that occur between:

a 0.3 and 0.5 b 0.25 and 0.7

c $\frac{1}{8}$ and 0.25 d 0.3 and $\frac{1}{3}$

30 The ratio of water to chemical concentrate in a garden spray is 4:7.

a If 800 mL of chemical is used, how much water must be added?

b If 140 mL of water is used, how much chemical concentrate is needed?

c If my bucket holds 5.5 litres, how much of each ingredient should I add to the bucket in order to fill it with the garden spray mix?

31 Simplify the ratios:

a 3:90 b 5:25 c 80:8

d $2\frac{1}{2}$:20 e $\frac{2}{5}$:12 f $\frac{4}{10}$:5

Additional learning unit

Copy the following algorithms into your workbook and solve them.

32 a $-5 \times 3 =$ b $-6 \times -7 =$

c $2 \times 8 =$ d $7 \times -3 =$

33 a $9 \div -3 =$ b $60 \div -12 =$

c $-24 \div 4 =$ d $-35 \div -5 =$

34 a $-5 + (-36 \div 6) \times 4 =$

b $(-6 + -9) \div 5 + 7 =$

c $24 - 6 + (-2 \times 7) =$

Assessment

Investigation

This assessment task gives you the chance to show what you understand about estimation, percentages, fractions, decimals and ratios in a real-life setting. The work you have done in the topic Gadgets will help you to complete this task.

Task 1: Practical investigation—group work

Imagine that one member of your group has been selected as the artist's model for a human statue that is to be erected in the centre of Port Moresby. The statue is to measure 46.5 metres from head to foot and will be in proportion to the student model.

Work together as a group to first estimate and then calculate the size of at least four different parts of the statue, such as length of thigh, arm and neck, or circumference of head and wrist. Make a table to show the comparison in sizes between the student model and the statue, and produce evidence of your calculations.

Your group will be assessed on your:

- ability to work productively as a group
- evidence of reasonable estimations made prior to calculations
- choice of appropriate problem-solving strategies
- correctness of calculations
- presentation of table of comparisons
- overall effort and persistence
- general presentation of final product
- ability to complete the task in the time allocated.

Total 40 marks

Assessment

Test

Task 2: Test of basic skills and routine applications

Ratios and rates

1 How much will 5 m of timber cost if it is advertised at the rate of K3 per metre?

2 How long will it take to travel 30 km at the rate of 6 km/h?

3 Oil is dripping from a container at the rate of 36 drops every 4 minutes. How long will it take to drip 300 drops?

4 Paint is being applied at the rate of 500 mL per 10 m^2. How much paint will it take to cover 25 m^2?

(4 marks)

Directed numbers

5 Below are details from the monthly bank statements for four students. Withdrawals are recorded as a negative transaction and deposits are shown as a positive transaction. Calculate the balance in each account at the end of the month.

Student A: +K56 + –K8 + –K3 + +K5 + –K8

Student B: +K5 + +K6 + –K15 + –K2 + +K4

Student C: +K1 + –K8 + +K23 + –K5

Student D: +K8 + –K15 + –K4 + +K7 + –K6

(4 marks)

6 Solve:

a $5 + 7 \div (-3 \times -5)$ b $36 - 4 - (2 \times -5)$ c $-14 + 120 \div -3$ (3 marks)

Area

7 a What is the diameter of a circle with a radius of 22 cm?

b What is the radius of a circle with a diameter of 30 cm?

c Estimate the circumference of a circle with a diameter of 8.5 m.

(3 marks)

8 Estimate the area of a circle with a 10 cm radius using the inscribed square method.

(4 marks)

9 Calculate the area of a circle with a diameter of 15 cm.

(2 marks)

Shapes

10 In your own words, explain why the symbol π represents the number 3.14 or $\frac{22}{7}$.

(3 marks)

Assessment

Test

Angles and shapes

11 Find the sum of the interior angles in the following polygons.

(4 marks)

a

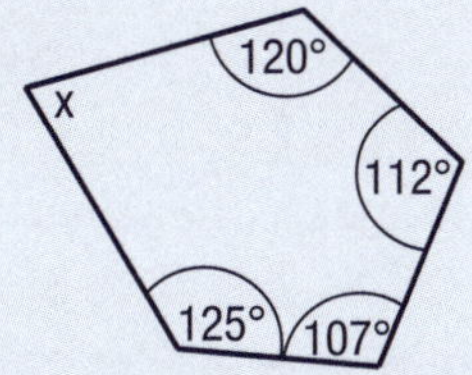

b

98° 98° 110° x 130° 110°

Angles

12 Calculate the size of the missing angle in each of the following examples. State whether the two adjacent angles are **complementary** or **supplementary.**

(4 marks)

a

x 28°

b

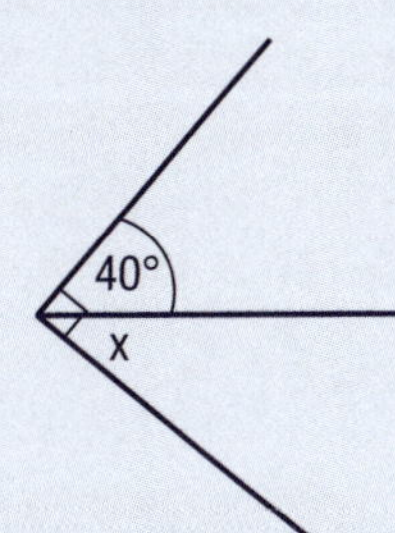

c

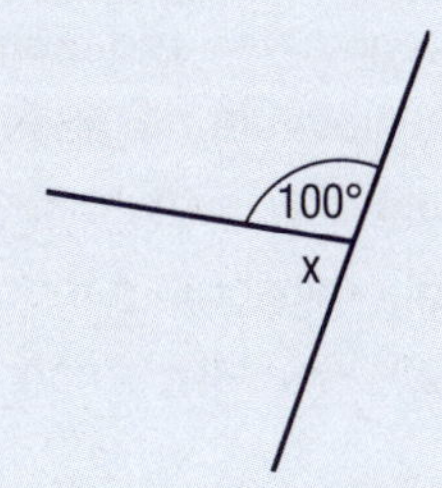

d

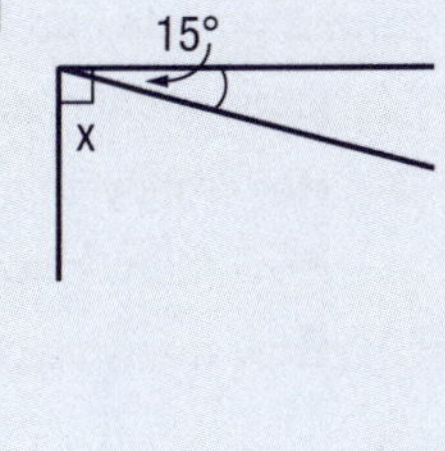

13 Find the size of the missing angles without using a protractor.

(5 marks)

a

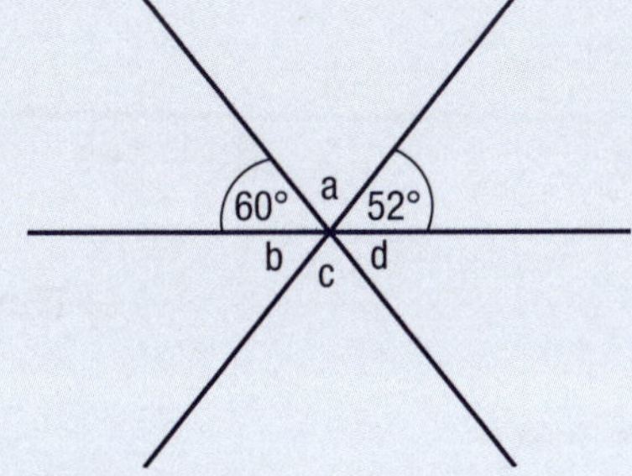

b

20° 24° a 100° e b c d

Assessment

Test

Time

14 A car was travelling at the rate of 80 km/h.

a What distance would the car travel in $3\frac{3}{4}$ hours at this rate?

b How long would it take a car to travel 420 km at this rate?

(2 marks)

15 A student earns K12.60 per hour in a part-time job.

a How many hours will the student need to work in order to earn at least K150?

b How much will the student earn for working $6\frac{1}{4}$ hours?

c Calculate the amount the student is being paid per minute.

(3 marks)

Sets

16 Eighty people at the market were surveyed about what they had purchased. Fifty people had bought vegetables, 32 had bought craft goods and 18 had bought tools. Three people had bought all three items and 2 people had bought vegetables and tools only.

a Draw a Venn diagram to illustrate the above information. (2 marks)

b How many people bought only craft items?

c How many people bought only craft goods and tools?

d How many people bought vegetables only?

(3 marks)

Algebra

17 Find the value of $3x + 2y$ if:

a $x = 12$ and $y = 10$ b $x = 8$ and $y = 15$ c $x = 30$ and $y = 12$

(3 marks)

18 If $p = 7$ and $k = 3$ calculate:

a $3p + k$ b $4(p - k)$ c $2p(8 - 3k)$ d $(10 + 5k)3$

e $p + 2(p - 2k)$

(5 marks)

19 Expand then simplify:

a $4 + 3(t + 5) - t + 4$ b $3(4m - 4) + 5(m + 3)$ c $5k(4 - 2h) + 4k$

d $h(4 - x) + 3h - 4x$ e $2(p - 5t + 6) + 2p$ f $2k(3 - 2k) + 5k$

(6 marks)

Topic 2 And the Winner is . . .

Learning Unit 1: The Games People Play

Learning Unit 2: Water Sports

Learning Unit 3: In the Bag

Learning Unit 4: Additional Learning, Revision and Assessment

Topic 2 And the Winner is . . .

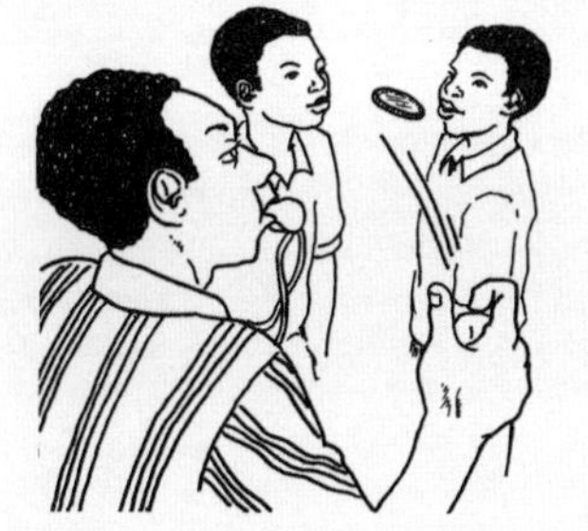

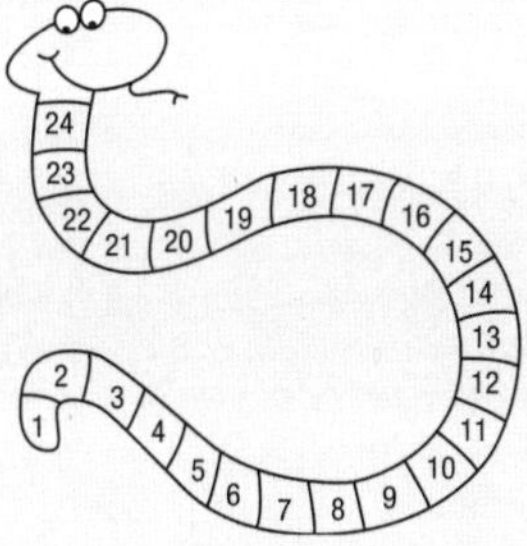

In this topic you will study the issue of gambling and, as well, investigate a range of indoor and outdoor games and sports.

The topic includes a study of chance in relation to gambling and life events generally. It also looks at different ways that certain sports are scored, and how packaging of equipment is related to the size and shape of the objects to be packaged.

The Assessment Tasks provide opportunity for you to demonstrate your understanding of the material covered both through project work and formal assessment.

Topic 2: *And the Winner is . . .* comprises the following Learning Units:

Learning Unit 1: The Games People Play

Learning Unit 2: Water Sports

Learning Unit 3: In the Bag

Learning Unit 4: Additional Learning, Revision and Assessment

Overview

Each Learning Unit reflects an aspect of mathematics used in everyday life. A brief summary of the mathematics in each Learning Unit appears below.

Learning Unit 1, *The Games People Play*, focuses on identifying the probability of particular outcomes. The issue of problem gambling is also addressed in the unit.

Learning Unit 2, *Water Sports*, explores the link between capacity and volume and examines the use, presentation and analysis of statistical information in real-life situations.

The work in **Learning Unit 3**, *In the Bag*, investigates the nets of solids and how to calculate the volume of various solids including prisms, cylinders, cones and pyramids. The unit also looks at problem solving using fractions.

Learning Unit 4, *Additional Learning, Revision and Assessment*, includes a stand-alone lesson about square and cubic numbers, a revision lesson, an investigation and a formal test.

Some of the materials that you may need to complete the activities in this topic are listed below.

You may need:

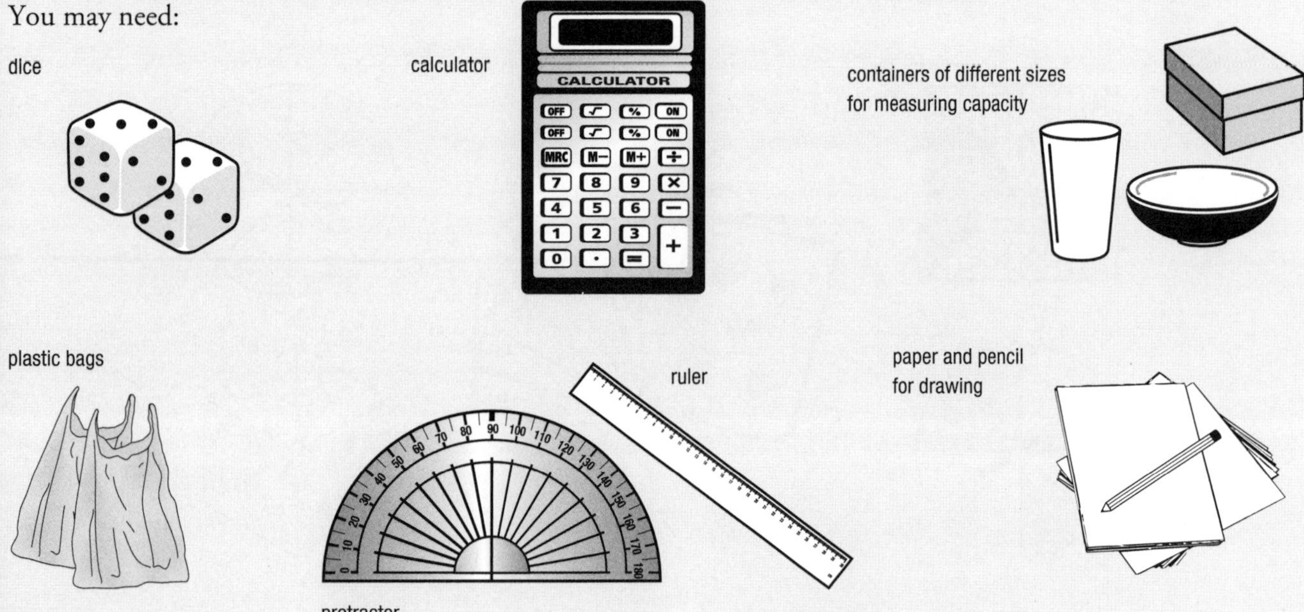

Learning Unit  The Games People Play

Strand: Number and Application

Ratios and Rates	Outcome 8.1.5	Apply ratios in solving problems from real life

Strand: Chance and Data

Probability	Outcome 8.4.3	Explore the social implications of probability
Estimation	Outcome 8.4.7	Estimate results of calculations

Strand: Patterns and Algebra

Algebra	Outcome 8.5.2	Recognise and use patterns in processes

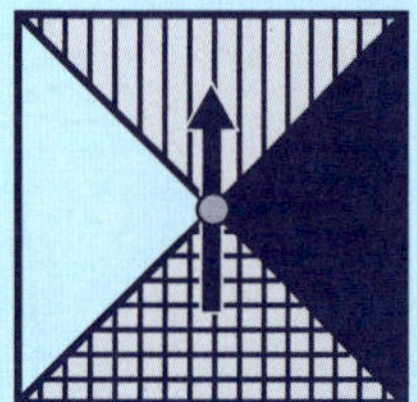

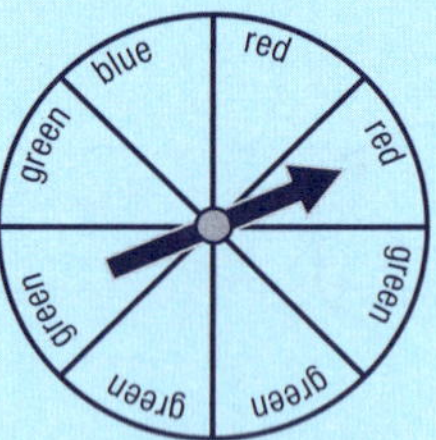

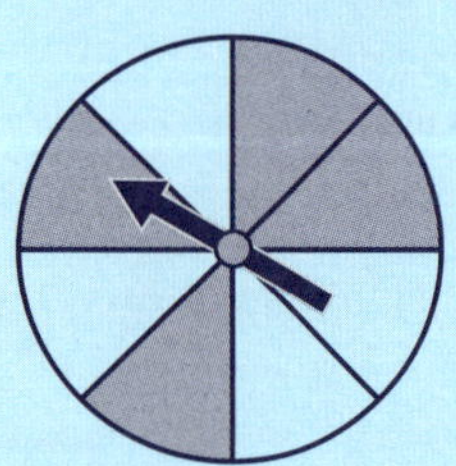

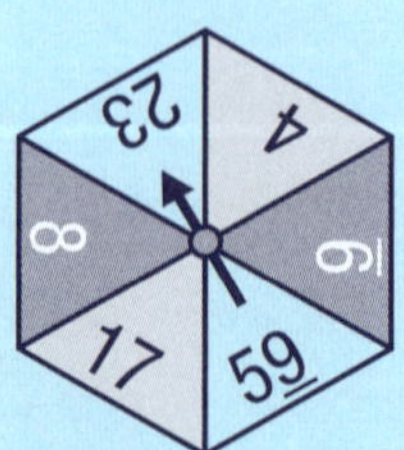

Lesson 1:	**Introduction**	Games of skill and games of chance Chance in everyday life
Lesson 2:	**What are my chances?**	Ratio Simplifying ratios
Lesson 3:	**Experimental probability**	Estimating the likelihood of an outcome Probability as long-run proportion
Lesson 4:	**Theoretical probability**	Using the formula to calculate probability
Lesson 5:	**Playing darts**	Scoring Calculating different ways of scoring points
Lesson 6:	**Other games—skill or chance?**	The game of hops Race to 21
Lesson 7:	**The problem with gambling**	The chances of losing Gambling profits the owner of the business
Lesson 8:	**Promotional gambling**	The social cost of gambling
Lesson 9:	**Some everyday risks and chances**	Chance in everyday situations

Lesson 1 Introduction

During this unit you will investigate how the characteristics of some games can influence the likelihood of a player winning. You will use your understanding of numbers and ratio, and the language of chance to describe the probability or likelihood of an outcome.

People sometimes gamble as part of playing certain games. While gambling can be a form of fun, for some people gambling becomes a problem. That happens when gambling begins to play too big a role in a person's life and the risks they take hurt themselves, their family, their friends and people who care for them.

Games are played by people all around the world. Some games (like chess) involve special skills and tactics to increase a player's chances of winning. Other games (like snakes and ladders) rely solely on chance to decide the winner. Most games require an element of mathematics. Knowing how to do the mathematics may help a player to win the game or at least be more aware of their chance of winning.

1 a Identify some of the games being played in these illustrations.

b Sort them into games of skill and games of chance.

2 List some of the games you have played in the past month.

3 Do you prefer games of skill or games of chance? Why?

4 List some other examples of both types of games.

5 Are games of skill or games of chance preferred by most students? Survey your class.

6 Which type of game do you have the most chance of winning—a game of chance or a game of skill? Explain your answer.

7 Everyday life is full of situations where people talk about 'chance'. Write a few sentences to describe how chance might be involved in each of the following examples:

a examination results

b job applications

c weather

d health

Lesson 2 What are my chances?

In Grade 7 you learned about ratio. Ratio can be used to think about what chance a person has of winning some games.

Ratio compares two or more different amounts.

For example, in this collection there are three triangles and two squares.

- The ratio of triangles to squares can be written as 3:2.
- The ratio of squares to triangles is 2:3.
- The ratio of triangles to the total number of shapes is 3:5.
- The ratio of squares to the total number of shapes is 2:5.

Bernadette put 10 items in a bag: 6 buttons, 3 coins and one pumpkin seed.

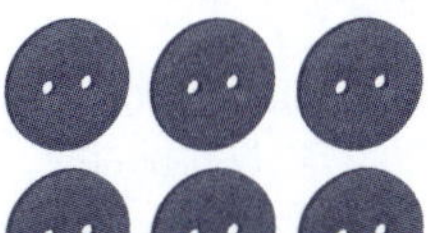

1 Write the ratio of:

- a buttons to coins
- b buttons to seeds
- c coins to total items in the bag
- d buttons to total items in the bag
- e coins to seeds.

2 Bernadette takes one item from the bag and asks Mary to guess what it is.

- a What is the item most likely to be? Explain why.
- b What item is it least likely to be? Explain why.

3

- a If Bernadette took one button and two coins out of the bag, describe the ratio now existing between the remaining items.
- b What item has the best chance of being taken out if they play the guessing game now?
- c What is the ratio of that item to the total number of items remaining in the bag?

Ratios can be simplified by dividing each number by the same value. This is done in the same way that we simplify fractions.

For example:

The ratio 15:50 can be simplified by dividing both numbers by 5.

So 15:50 can be simplified to 3:10.

The ratio 32:28 can be simplified by dividing both numbers by 4.

So 32:28 can be simplified to 8:7.

In a similar way, if we multiply both numbers in the ratio 2:7 by 9, we find that the ratio 18:63 has the same value as the ratio 2:7.

4 Simplify the following ratios:

a	6:15	b	10:48	c	88:16
d	33: 6	e	125:25	f	15:85

5 Harold has 6 picture cards. Four are coloured and the rest are black-and-white.

- a What is the ratio of coloured cards to black-and-white cards?
- b If he increases the pack to 12 cards and uses the same ratio, how many black-and-white cards will he have?

c How many black-and-white cards will he have if he increases the pack to 30 cards with the same ratio of coloured cards to black-and-white cards?

d If he shuffles the 30 cards and draws one card at random, is he more likely to draw a coloured card or a black-and-white-card? Explain your answer.

6 Copy the following game board into your book and use three different coloured pencils to show a ratio of 3:2:1 on the squares.

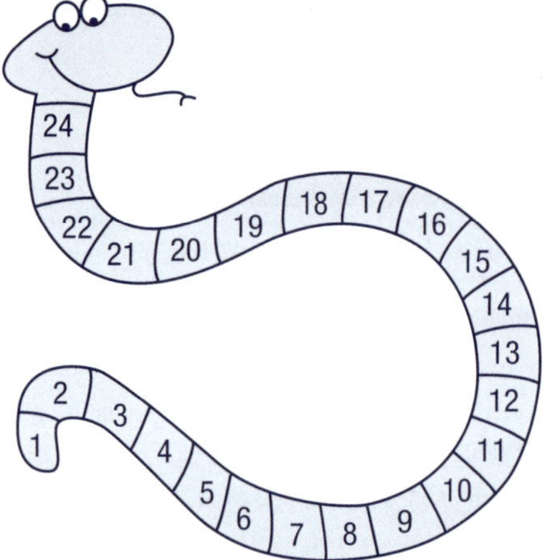

Challenge

Create a game that could be played using your game board. Write the rules and test the game with a friend. Make improvements to the rules if needed.

7 Harold and Fegsley created some games for two people to play using one dice. Work with a partner to answer the following questions about their games.

a Roll the dice. Player 1 wins if a 6 is rolled; Player 2 wins if any number other than 6 is rolled. Is this a fair game? Explain why.

b Roll the dice. Player 1 wins if an even number is rolled; Player 2 wins if an odd number is rolled. Is this a fair game? Explain why.

c Roll the dice. Player 1 wins if a number above 4 is rolled; Player 2 wins if a number below 4 is rolled. Is this a fair game? Explain why.

d Roll the dice. Player 1 wins if number 1, 5 or 6 is rolled; Player 2 wins if number 2, 3 or 4 is rolled. Is this a fair game? Explain why.

e Roll the dice. Player 1 wins if a multiple of 2 is rolled; Player 2 wins if a multiple of 2 is not rolled. Is this a fair game? Explain why.

f Roll the dice. Player 1 wins if a 3 is rolled; Player 2 wins if a multiple of 5 is rolled. If any other number is rolled, they roll again until a 3 or a multiple of 5 is rolled. Is this a fair game? Explain why.

8 Rodney and Sam play the following game. Roll the dice. Rodney wins if an even number is rolled. Sam wins if an odd number is rolled.

a If they play the game 100 times, how many times would Rodney probably win? Explain your answer.

b If they play the game 150 times, how many times would Sam probably win? Explain your answer.

c If they played the game many times and Sam won 75 times, how many games would Rodney probably have won? Explain your answer.

9 Jane and Betty play the following game. Roll the dice. Jane wins if the number 1 or 3 is rolled. Betty wins if any other number is rolled.

a If they play the game 100 times, how many times would Jane probably win? Explain your answer.

b If they play the game 120 times, how many times would Betty probably win? Explain your answer.

c If they played the game many times and Jane won 40 times, how many games would Betty probably have won? Explain your answer.

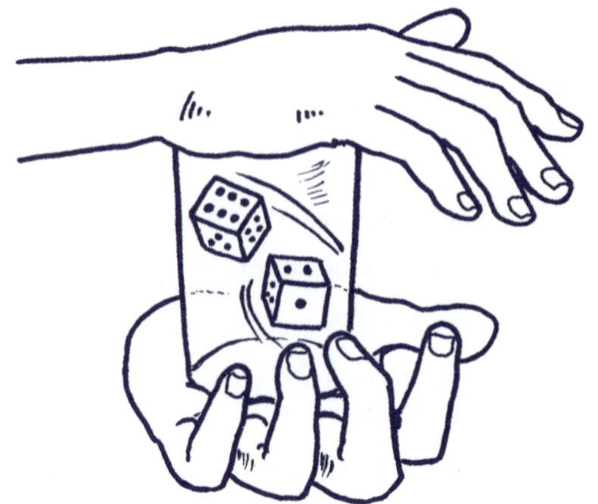

Lesson 3 Experimental probability

We hear statements of probability every day. For example, 'My team will probably win today', 'I've got a 50-50 chance of getting the job', 'I'm certain to pass this exam', 'It's very unlikely he'll win the race', 'I'll never live to be 110 years old'.

Probability means predicting the likelihood that something will or will not happen. We can collect statistical data to help us calculate the probability of a particular outcome occurring.

If I drop a drawing pin, it has a chance of landing point up or point down.

point up point down

Some students collected data from 200 pin drops. Their results are in this table.

Point up	Point down
94	106

They calculated the probability of the pin falling point up as being 0.47. This is almost a 50:50 chance. Probability can be written as a fraction, a decimal, a percentage or a ratio.

Help Box

The 'pin' experiment calculates probability in relation to the proportion of occurrences 'in the long run'. This literally means the likelihood of something occurring in a large number of experimental trials. The experimental probability is calculated by writing a fraction to represent the total number of 'point up' scores in relation to the total number of throws.

The fraction in the example above can be written as: $\frac{\text{point up total}}{\text{total number of throws}}$ OR $\frac{94}{200}$.

The experimental probability of a 'pin up' result was $\frac{94}{200}$ or 0.47 or 47% or 94:200.

1 a Estimate how many times a card that is thrown into the air 100 times will land 'playing side up'. Give an explanation for your answer.

b Write your estimation as a decimal using the formula for calculating experimental probability.

2 a Work with a partner to take turns throwing one playing card into the air and make a table to show the number of times the card lands 'face down'.

b Calculate the experimental probability of the card landing 'face down' based on your table of results.

3 Calculate any difference between your estimation in question 1 and what actually happened in question 2. Write a few sentences to explain how this could occur.

4 a Estimate how many times a normal dice that is rolled 50 times will land on six. Give an explanation for your answer.

b Write your estimation as a decimal using the formula for calculating experimental probability.

5 a Work with a partner to take turns rolling a dice and make a table to show the number of times a 6 is thrown.

b Calculate the experimental probability of rolling a 6 based on your table of results.

6 Calculate any difference between your estimation in question 4 and what actually happened in question 5. Write a few sentences to explain how this could occur.

7 The table below shows the results some students obtained when spinning a spinner with four equal sectors for a total of 200 spins.

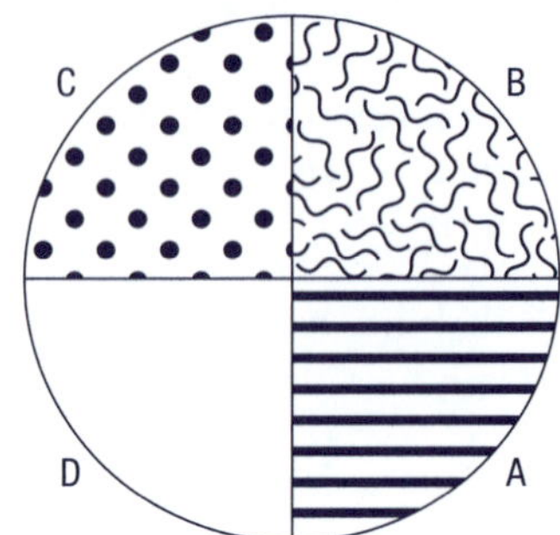

Pattern	A	B	C	D
Number of occurrences	45	55	40	60

a Calculate the experimental probability of each sector being spun and write your answers as decimals.

b What would you predict the probability of each pattern being spun might be if the experiment was trialled for 1000 spins instead of 200? Explain your answer.

Challenge

Design an experiment involving at least three possible outcomes that could be used to test probability. Estimate the probability of each outcome and then complete the experiment to see how close your estimation was.

Lesson 4 Theoretical probability

In the previous lesson we calculated experimental probability by collecting data and calculating probability as long-run proportion. When a situation has equally likely outcomes, it is possible to estimate the probability of an event without doing an experiment. Estimating probability in this manner is called 'theoretical probability'.

Probabilities are usually written as fractions or decimals between 0 and 1. A probability of 0 means that it is impossible for the event to occur. A probability of 1 means the event is certain to occur.

1 Copy the number line into your book. Mark the line to show where each of the following descriptions would be located.

a	very likely	b	little chance	c	uncertain	d	possible
e	equal chance	f	not likely	g	no hope	h	sure

2 Use one of the following terms to describe each of these statements: impossible, unlikely, likely, certain.

a Christmas Day being 25 December

b throwing a 'head' when you toss a coin

c you going to Australia in the next two years

d snow falling in Port Moresby in January

e rolling a seven on a standard dice

f drawing an ace from a normal pack of playing cards

g rain falling in the school grounds today

h the sun rising tomorrow

i Papua New Guinea winning a gold medal at the next Olympic games

j your soccer team winning the season premiership

3 Write a statement of your own to describe a situation that is:

a impossible b unlikely c even chance d likely e certain

Help Box

We can calculate theoretical probability by dividing the number of favourable outcomes by the total number of possible outcomes.

$$\text{Probability (Pr)} = \frac{\text{number of favourable outcomes}}{\text{total number of possible outcomes}}$$

For example, the probability of throwing a 6 on a normal dice is Pr $\frac{1}{6}$. This can also be written as Pr 0.17 or Pr 17% or Pr 1:6 as they are all equivalent.

4 Copy each of the following statements into your book and choose the answer that you think is closest to the correct response in each case. (Note: there are 52 cards in a pack of playing cards.)

a	The probability of tossing a coin and getting a head is:	$\frac{1}{2}$	1	$\frac{3}{8}$	$\frac{1}{3}$
b	The probability of drawing a queen from a pack of playing cards is:	$\frac{1}{4}$	$\frac{2}{3}$	0	$\frac{1}{13}$
c	The probability of rain during the next month is:	0	0.2	1	0.7
d	The probability of rolling an even number on a standard dice is:	0.1	0.6	1	0.5
e	The probability of drawing a heart from a pack of playing cards is:	4%	52%	25%	75%
f	The probability of a maximum temperature tomorrow of 10°C is:	10%	50%	20%	80%

5 Use the formula to calculate the theoretical probability of the following chance events:

a drawing a diamond from a pack of well shuffled playing cards

b drawing an ace from a shuffled pack

c drawing a red card from a shuffled pack

d drawing a number card from a shuffled pack

e rolling a 3 with a 6-sided dice

f rolling an even number with a 10-sided dice

g rolling a multiple of 3 with a 10-sided dice

Challenge

Calculate the theoretical probability of scoring 8 when rolling two six-sided dice and totalling their scores.

6 A spinning wheel has the numbers 1 to 30 evenly spaced around it. Only 30 tickets are sold each spin. What is the theoretical probability that Edna will win if she has bought:

a one ticket only?

b all the even numbered tickets less than 20?

c all the tickets with a digit 4 in them?

d her favourite ticket numbers 2, 5, 13, 20 and 25?

7 This spinner has the probability of $\frac{1}{4}$ of landing on the grey shaded section.

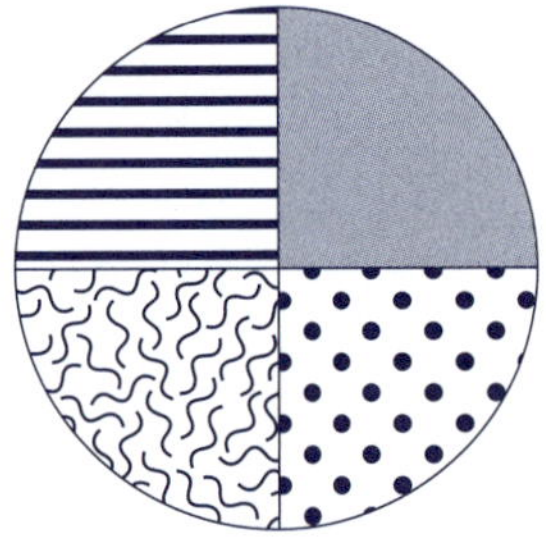

Calculate the probability of the following spinners landing on a grey shaded area.

a b

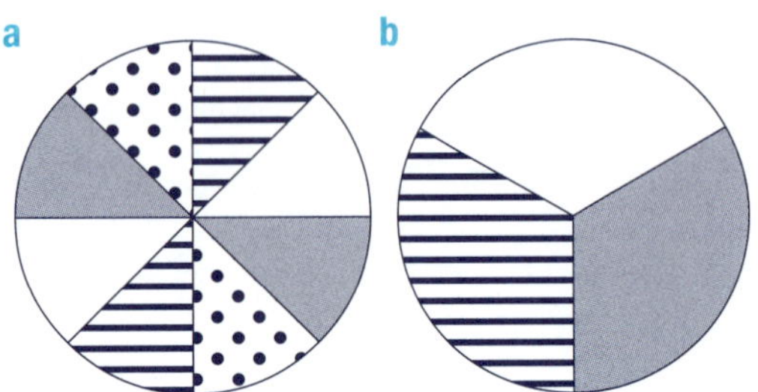

c d

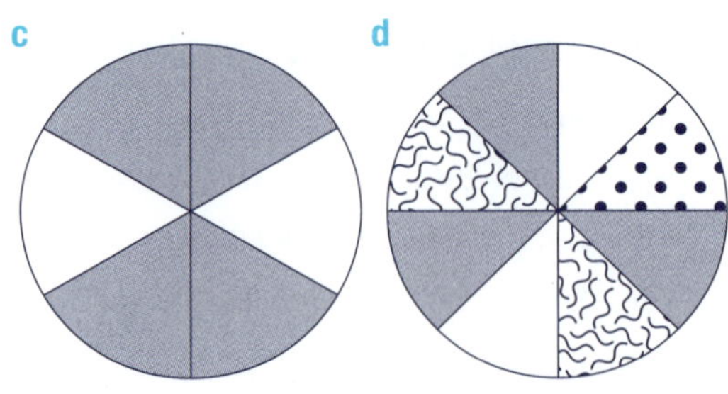

8 Draw a spinner to show each of the following probabilities:

a probability of landing on a white section $\frac{3}{8}$

b probability of landing on a striped pattern $\frac{1}{3}$

c probability of landing on a white section $\frac{1}{3}$ and a striped section $\frac{1}{2}$

d probability of landing on a shaded section 0.25 and a spotted section 0.5

e probability of landing on a spotted section 0.2, a black section 0.3 and a striped section 0.3

Challenge

Which of the probabilities described in question 8 above has the same theoretical probability as selecting a circle from this set of objects?

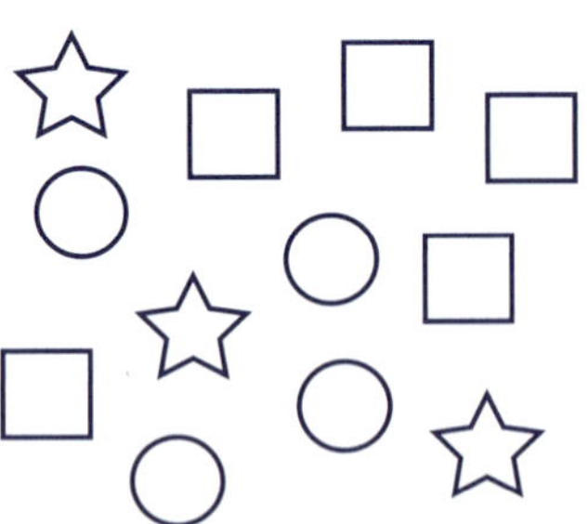

Lesson 5 Playing darts

The object of a game of darts is to score the most points in a round or in an entire game. A dartboard comprises 82 sections. A dart landing on one of the 20 numbered sections scores that number or its double or triple value.

Players stand two metres from the board to throw. Each player has three darts per turn and the player with the highest number wins the round.

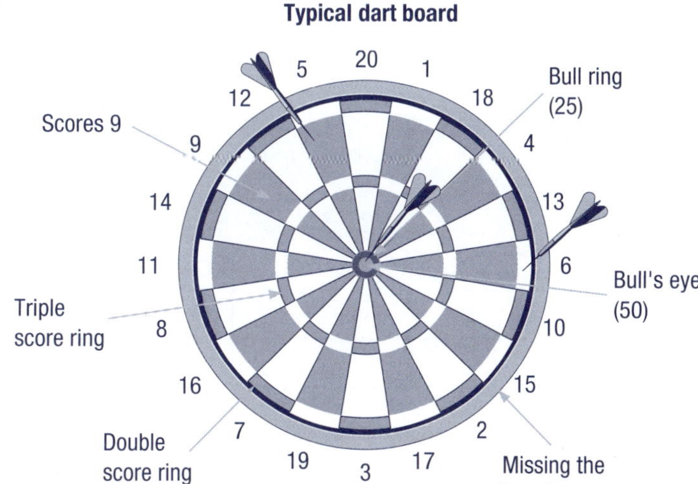

1 Assuming the three darts hit the board, calculate:

- **a** the highest score possible in any round
- **b** the lowest score possible in any round
- **c** the total score for the three darts in the diagram

2 Suggest three locations on the dartboard that a player could throw to beat the score shown in the diagram above.

3 For each of the following, suggest a combination of scores you would have to throw to win if your opponent had thrown:

- **a** a seven, a ten and a one
- **b** a bull's eye, a double twenty and a single ten
- **c** a triple eighteen, a double ten and a six
- **d** two bull rings and a bull's eye

4 Where possible, list at least three different ways in a single round that someone can score:

- **a** 125
- **b** 150
- **c** 70
- **d** 5
- **e** 8
- **f** 13
- **g** 25
- **h** 3
- **i** 21
- **j** 24

5 Discuss with a friend the strategy you would use to win each round if you were playing a game of darts.

- **a** Write down your strategy, explaining why you think it would work.
- **b** If possible, play a game of darts to test your strategy.

6 Think about the amount of skill, mathematics and chance involved in playing darts. Write a sentence to describe how important each of these factors is in determining who wins.

Lesson 6 Other games—skill or chance?

Hops

The object of the game is to be first to score 120 (or more). Two normal dice are used, as well as a third with stickers on the faces. The stickers show:

+	−	×	÷	RA	RA

RA means roll again until one of the four operation symbols is rolled.

All three dice are rolled together by Player 1. The player uses the two numbers and the operation symbol to create an algorithm and calculate the answer. The score is recorded and the dice are given to the next player on the right to repeat the process and so on.

When all players have had a turn, the player with the highest score becomes Player 1 and the next round continues. Play continues until a player scores 120 and calls 'Hops'.

1 Copy the table into your book and use it to record the round scores as you play a game of Hops with three other students.

	PLAYER 1		PLAYER 2		PLAYER 3		PLAYER 4	
	Round score	Running score	Round score	Running score	Round score	Running score	Round score	Running score
Round 1								
Round 2								
Round 3								
Round 4								

2 Discuss with a friend then write a sentence to describe what you think is:

a the best operation to roll

b the worst operation to roll

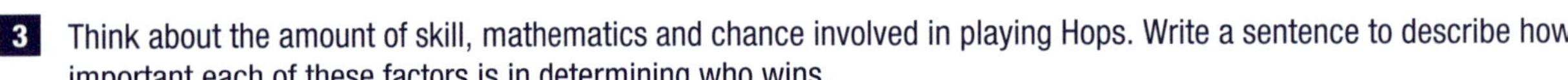

3 Think about the amount of skill, mathematics and chance involved in playing Hops. Write a sentence to describe how important each of these factors is in determining who wins.

Race to 21

The object of this game is to be the first person to say 'twenty-one'. The game is played by two people. One player begins by saying either the number 1 or 2. The other player then has the choice of adding either one or two to that number and says the new number. Player 1 then has the choice of adding either one or two to the number said by Player 2 and so on, until a player has the opportunity to say 'twenty-one'.

4 Play a game of Race to 21 with a friend.

5 Which number closest to 21 does a player need to say to ensure that they will win?

6 Play the game again and test your theory.

7 What other numbers might be critical for a player to say if they want to win? Write a few sentences to describe your thinking.

8 Think about the amount of skill, mathematics and chance involved in playing Race to 21. Write a sentence to describe how important each of these factors is in determining who wins.

Challenge

What are the critical numbers if the rules for Race to 21 are changed to allow players to begin with 1, 2 or 3 and to add either 1, 2 or 3 to the previous player's number?

Lesson 7 The problem with gambling

Many adults in Papua New Guinea gamble, whether at cards, the horses, sports events, lotteries or some other game of chance. Gambling can be a bit of light-hearted entertainment. The key to being a responsible gambler is to stay in control and remember that gambling is just a game: it is a form of entertainment that costs money. Even if you win once in a while, gambling will cost you more to play than you will ever be likely to receive. The more you play, the more you pay.

What are the odds of winning?

The odds of winning differ for different types of gambling activities. Commercial gambling systems are not designed to provide the player with extra income, rather they are designed to make a profit for the owner of the business.

Help Box

The sum of the probabilities of all possible outcomes of a simple event is 1.

For example, if a normal dice is thrown there are six possible outcomes of which throwing a 4 is one outcome. (Pr of being a 4 is $\frac{1}{6}$).

There are five possible outcomes that are not 4. (Pr of not being a 4 is $\frac{5}{6}$).

The two events are *complementary*. This means they combine to equal 1.

A person is asked to guess which of the three coconut shells hides the token.

The probability of selecting the correct coconut and winning the game is Pr $\frac{1}{3}$. However, the probability of selecting the wrong coconut shell is greater at Pr $\frac{2}{3}$.

A

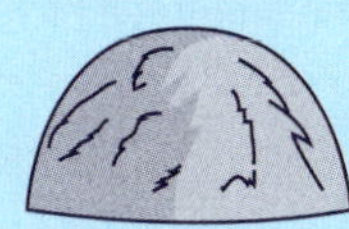

B

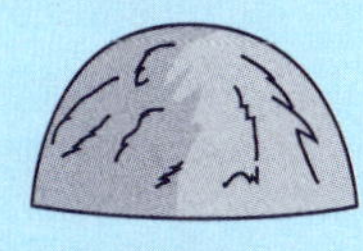
C

1 At the fair, the stallholder sells 200 tickets to raffle a chicken. Harold buys 5 tickets in the raffle.

a What is the probability that he will win?

b What is the probability that he will not win?

2 The lottery has 10 000 tickets. Betty buys 10 tickets.

- **a** What is the probability that she will win first prize?
- **b** What is the probability that she will not win first prize?
- **c** Betty's mother has a probability of $\frac{1}{100}$ of winning the first prize. What is the probability that she will not win first prize?
- **d** There are other prizes in the lottery. Betty has a probability of $\frac{7}{500}$ of winning any one of the lottery prizes. What is the probability that she will not win a prize?

3 At a school fair people pay 50 toea for a chance to draw a marble from a bag (without looking). The bag contains marbles labelled 1, 2 or 3, and there are equal numbers of marbles with each number. If a marble labelled 1 is drawn out, the person wins K1. There is no prize if either of the other numbers are drawn.

- **a** What is the Pr of a person winning K1 if they draw from the bag once?
- **b** What is the Pr that they will not win?
- **c** What is the Pr that the stallholder will get to keep the person's 50 toea and not have to pay them K1?

Challenge

If 60 people each paid to play this game, calculate the profit that the stallholder would make based on the probability of people withdrawing a winning marble.

4 The sign to the right was above a market stall.

- **a** Copy the table below into your book. Complete it to show the full range of possible outcomes when rolling two dice and totalling the sum of the two numbers shown. The first three have been done for you.

Sum of two dice	2	3	4	5	6	7	8	9	10	11	12
Combinations	1, 1	1, 2	1, 3 2, 2								

- **b** Which number is the most likely to be scored in this game?
- **c** The stall owner says people will win a prize if they roll any of the eight numbers listed on the sign. At first glance it appears that this will be an easy game to win. Write a few sentences to describe why the stall owner will probably not have to give away many prizes in this game.

Two dice roll

Score a total of
2, 3, 4, 5, 9, 10, 11 or 12
and WIN!

5 At a fundraiser for the local sports club, people were playing this game. A player flips three 20 toea coins above the board. If a coin lands on a line the player loses the coin. If a coin lands on a winning space, the player wins a coin of the marked value as well as getting their original 20 toea returned.

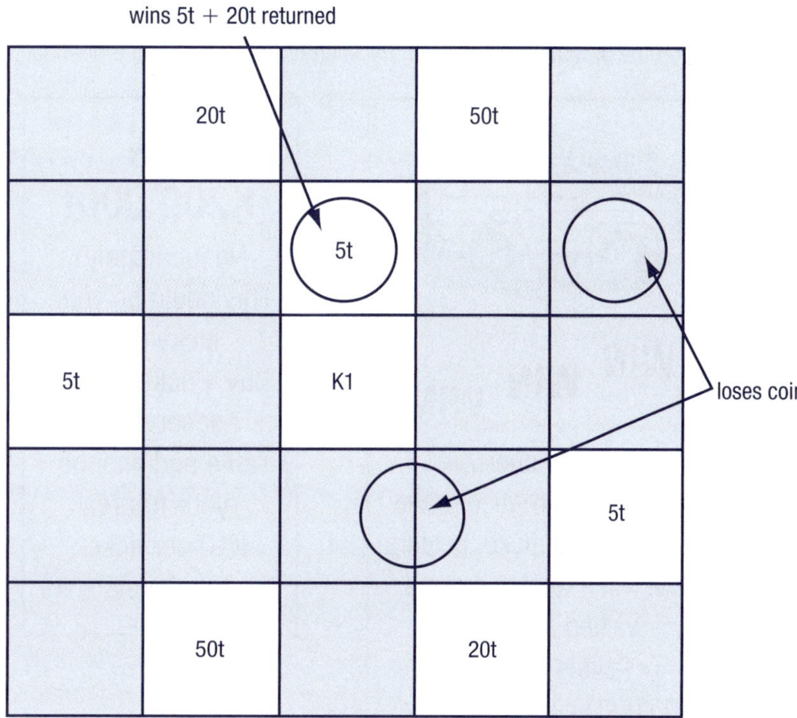

a Do you think this is a fair game? Why?

b Write a few sentences to predict whether the club is likely to make money or lose money from this game. Explain the reasons for your answer.

6 a Play the game described above with a partner. One of you could represent the club, while the other represents ten different people who play the game. Flip the three coins from about 20 cm above the grid to avoid them missing the grid completely. Keep a record of how much money the club won and how much money each person won over the 10 games.

b How close were your predictions?

c How successful would the game be as a fundraiser for the club?

d What changes might need to be made to the game to improve its potential? Explain your answer.

Lesson 8 Promotional gambling

A report by the National Council of Churches in Australia (http://www.ncca.org.au/home_page) suggests that 15% of regular gamblers are susceptible to problem gambling. This means that their gambling habit is likely to cause damage to themselves or others.

1 a List at least three different forms of gambling that you know occur within Papua New Guinea.

b Rate the three types of gambling from the one you think is most popular to the one you think is least popular.

c Explain why you think the first item on your list is the most popular.

The report mentioned above also describes some concerns about advertisements aimed at encouraging people to gamble. These include:

- Advertisements emphasise the chances of winning without showing the chances of losing.
- Gamblers are shown in ideal terms as young, attractive and happy. This, however, is not the norm.
- Advertisements may target some people's belief in fate and chance.

2 Use your knowledge of probability to create an advertisement to show a more realistic idea including the probability of losing alongside the probability of winning for each of the following promotional advertisements.

a

WIN WIN WIN

1 lucky winner will
win a luxury 4-wheel-drive.
99 other lucky ticket holders
will win a total prize pool
valued at K5000.
Tickets K3 each.
100 000 tickets to be sold.

b

WIN
K200 000!

No luck lately?
This could be your
lucky day!
Buy a ticket in the
'I deserve this'
raffle and change
your future.
K10 per ticket.
50 000 tickets sold.

c

NEED A HOLIDAY?

You could be
1 of 4 happy couples
jetting off to
Sydney, Australia,
for 2 glorious weeks.
The 'getaway' raffle.
Tickets K10 each.
5000 tickets offered.

3 While at the supermarket, Francis heard about the following raffle.

Don't cook tonight!

Buy a K2 raffle ticket
and **WIN**

- a large roast chicken
- 2 large serves of chips
- 1.25 L cola
- 1 large coleslaw

One lucky customer
drawn each hour.
Winners must be in the store
to claim their prize.

He thought he would surprise his family with a free dinner, so he bought a ticket.

a What things might impact on Francis's chance of winning the raffle?

b Use your estimation skills and your knowledge of probability to predict the likelihood of Francis winning the raffle. Explain your answer.

4 The Crunchy Breakfast Company advertises that for a limited time each of their 500 g packets of cereal will contain a collectable card featuring an outstanding member of the National Rugby League. There are 25 cards in the series and a 500 g packet of the cereal costs K2.20.

a What is the minimum amount a family will have to pay to collect the entire series of cards?

b Each carton of 25 boxes of cereal contains one entire card series. If someone wanted a particular card and they bought the first box from a carton, what is the probability that the card they want will be in their box of cereal?

c Explain how this might change if one packet of cereal has already been sold from the carton before the person buys a packet.

Challenge

If a person had been given one card from the series already, what might be the probability of them getting a different card if they purchased the first box from a new carton? Discuss the question with your friend and write a few sentences to describe your thinking.

5 The 'Crunchy Breakfast Company' decides to increase their carton size to 55 boxes of cereal and reduce the number of cards in the series to 20. The marketing team suggests that each carton should now contain:

Player number	1	2	3	4	5–20	TOTAL cards per carton
No. of cards per carton	1	2	2	2	3	
Total cards	1	2	2	2	48	55

What is the probability that a box from a new carton will contain a card about:

a Player number 1? b Player number 2? c Player number 4? d Player number 20?

6 Explain why the marketing team's idea might help boost the company's sales.

Lesson 9 Some everyday risks and chances

Many commercial gambling outlets are organised so that a percentage (for example, 85%) of the money wagered is paid back to customers in winnings, and the remainder (15%) is kept as profit for the business. Not everyone wins. Those who do, gain at the expense of others who lose. The following case study describes how impulsive gambling can have long-term consequences.

Wesley has K20, the amount he needs to buy new pants to wear to a job interview tomorrow. He wishes he had enough to buy a new shirt also. Wesley visits the local casino hoping to double his money. He bets K5 on 'lucky six' in a dice game. He loses. Wesley bets K5 on six again. Number 2 is rolled. Wesley is sure six is lucky. He puts K5 on six again and loses again. Angry now, Wesley puts K2 on 'lucky six'. He wins his K2 back, plus K1. Wesley decides to leave the game and go home.

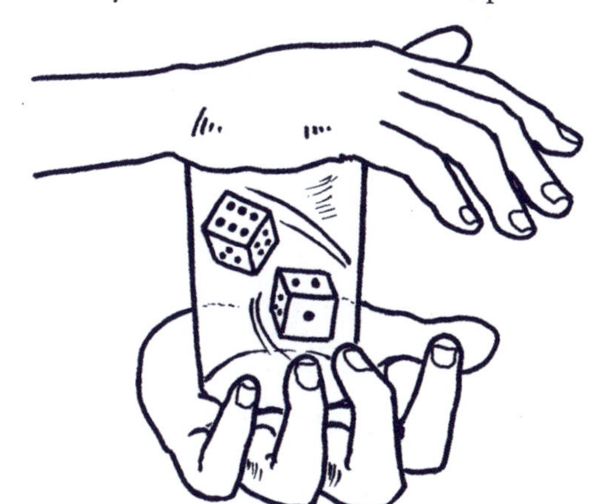

1 a How much did Wesley place in bets?

b How much did he win back?

c How much did he have altogether by the end of this story?

2 Will Wesley be able to buy the pants and the shirt with that amount of money?

3 Write a few sentences to explain to Wesley why his understanding of six being lucky is incorrect.

4 What fraction of the total amount that Wesley placed on bets was returned to Wesley in winnings?

5 If Wesley decides to gamble the entire K20, based on the 85% return example what amount of money could Wesley expect to get back?

6 Thinking about the reason Wesley wanted the extra money, what might be a long-term consequence of Wesley's gambling today?

A fast food chain has decided to open shops in certain areas. Each outlet will bring 150 new jobs to the area. The table shows an estimate of the number of unemployed people over 18 years old seeking work in each of the targeted areas.

	National Capital District	Western Highlands	Madang	Morobe
Male	15 200	3 000	2 700	6 000
Female	11 800	3 000	2 500	5 500
Total	27 000	6 000	5 200	11 500

(Source: adapted from 2000 National Census Report).

7 a Which area has the largest number of unemployed adults?

b In which area is there an equal likelihood of both men and women being unemployed?

c In which areas is there less chance of being unemployed if you are a woman than a man?

8 Calculate the different probabilities of being one of the 150 people to be employed by this company if you were unemployed in each of the areas listed.

9 In which area does an unemployed person have the best chance of being employed by the company? Explain your answer.

10 If only females applied for the jobs with the company, the unemployed women in which of these areas would have the best chance of being employed? Explain your answer by comparing probability.

This map shows where earthquakes have occurred in Papua New Guinea during the past 20 years.

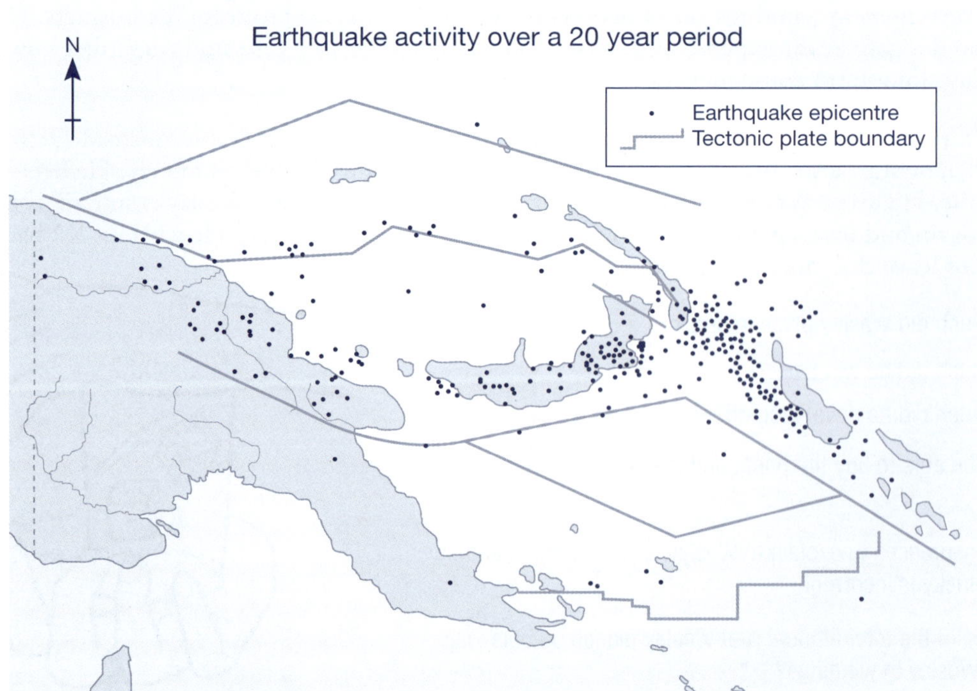

11 Use your estimation skills and knowledge of probability to suggest the probability of earthquake activity during the next 20 years in:

a Port Moresby

b Morobe

c Pomio

d Daru

12 A tray of seedlings has a $\frac{5}{8}$ chance of surviving the drought.

a List some things a farmer could do to increase the probability of the seedlings surviving.

b Select one of the suggestions on your list and estimate what the seedlings probability of surviving would become if the farmer acted on your suggestion.

13 Eating sensibly and getting regular exercise can increase a person's chances of staying healthy.

a List three other things people can do to increase their chances of staying healthy.

b Organise your list in order from the thing that would have the greatest influence on your chances of staying healthy to that with least influence.

Learning Unit  Water Sports

Strand: Number and Application

Decimals	Outcome 8.1.2	Use decimals to solve real life problems

Strand: Space and Shape

Capacity	Outcome 8.2.7	Apply capacity and volume measurements in problem solving
Capacity	Outcome 8.2.8	Convert between a variety of units: capacity and volume, and metric and imperial

Strand: Chance and Data

Statistics	Outcome 8.4.1	Interpret information presented statistically
Error and Accuracy	Outcome 8.4.4	Combine error-reducing strategies to get the best results

Lesson 1: Introduction	Water sports in Papua New Guinea Collecting data about student swimming ability
Lesson 2: Capacity and volume	Calculating volume Determining capacity from volume Determining volume from capacity
Lesson 3: Problem solving	Using capacity and volume in everyday situations
Lesson 4: Conversions	Converting the imperial and metric systems Working with capacity and volume
Lesson 5: The maths of diving	Using decimals in scoring Calculating the mean
Lesson 6: Swimming sensations	Comparing swim-times to a 100th of a second Calculating the mean
Lesson 7: Percentage change	Increasing and decreasing numbers by percentages
Lesson 8: Statistical average	Calculating the mean, median and mode
Lesson 9: Data displays	Drawing graphs and tables Interpreting data
Lesson 10: Misuse of statistical data	Analysing inappropriate displays of data

Lesson 1 Introduction

In this unit you will revise your knowledge about volume and capacity and learn how to convert between metric and imperial measurements when solving problems. You will also interpret statistical data about some popular water sports of Papua New Guinea.

Papua New Guinea is world-renowned for the fishing, diving, boating and cruising opportunities that it offers both locals and tourists alike. The repeated medal successes of Ryan Pini, one of Papua New Guinea's swimming sensations, have further helped to make water sports and Papua New Guinea seem synonymous.

1 List some of the water sports you know.

2 Which ones have you tried?

3 Which ones do you hope to try in the future?

4 What do you know about Ryan Pini?

Ryan Pini has swum the following events:

- 50-metre and 100-metre breaststroke
- 50-metre and 100-metre butterfly
- 100-metre backstroke
- 50-metre and 100-metre freestyle

5 Survey the class to determine how many students can swim those distances for each swimming style. Collect separate data for males and females.

6 Create graphs to show your results.

7 Write a short report to explain your findings.

8 Comment on how representative your class might be compared to other Grade 8 classes in other parts of Papua New Guinea.

Lesson 2 Capacity and volume

In Grade 7 you learned that 'volume' is the measure of space taken up by a three-dimensional object, and that the space inside a container is known as its 'capacity'. Although volume is calculated by measuring the dimensions of the object and capacity relates to measures of liquids, the strong link between the two concepts means that they are often used interchangeably.

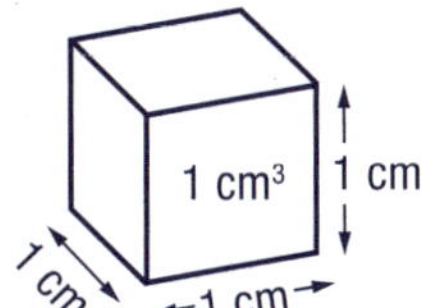

A container that holds 100 mL has a volume of 100 cm³.

Two litres of liquid has a volume of 2000 cm³.

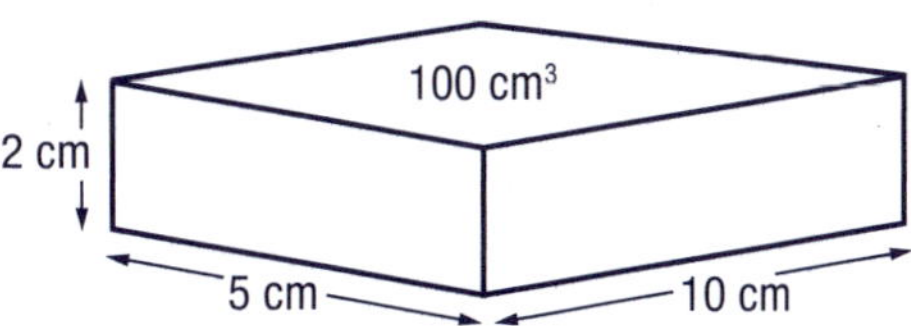

1 Use the formula $l \times w \times h = V$ to calculate the volume of these rectangular prisms.

a

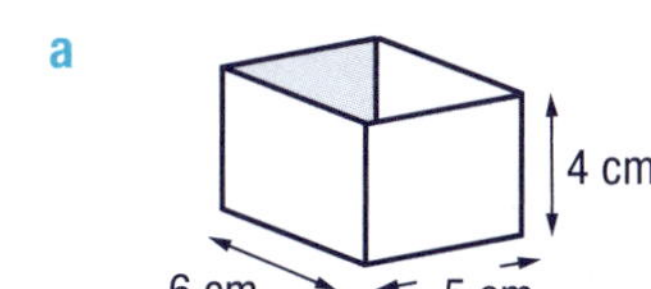

b

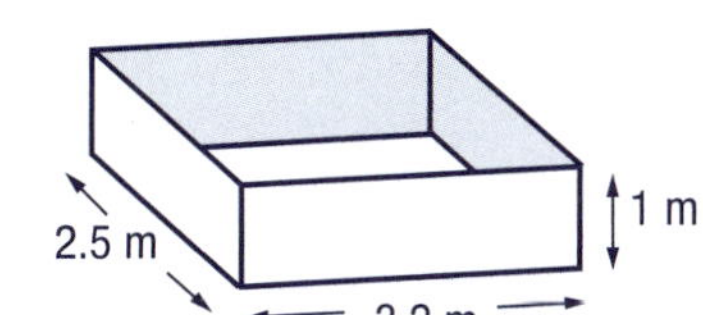

c

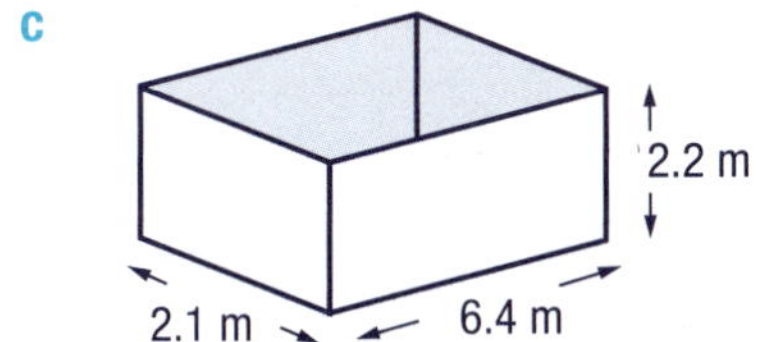

2 Copy the following table into your workbook and complete it.

	Dimensions			Volume (cm³)	Capacity (mL)
	Length	Width	Height		
	3 cm	3 cm	3 cm		
	12 cm	2 cm	2 cm		
	10 cm	5 cm	4 cm		
	15 cm	1 cm	0.5 cm		
	4 cm	2 cm	12 cm		
	8 cm	6 cm	5 cm		

3 Calculate the volume of the following rectangular prisms:

a width 15 cm, length 10 cm and height 20 cm

b length and width both 25 cm and 12 cm tall

c a cube with side lengths of 4 m

d height 3 m and a base area of 16 m²

4 Using your answers to question 3, calculate the capacity of each of the containers described.

5 Calculate the capacity of containers with the following volumes:

a $50\ cm^3$ b $85\ cm^3$
c $1250\ m^3$ d $2.5\ m^3$

6 Calculate the volume of containers with the following capacities:

a 20 mL b 330 mL c 8 L d 3.25 L

When an object is submerged in a container of water, the amount of water it displaces is equal to the object's volume.

7 If the following amount of water is displaced, what is each object's volume?

a 30 mL b 250 mL c 56 mL
d 375 mL e 2 L f 3.75 L
g 2300 mL h 1780 mL

8 How much water would objects of the following volumes displace?

a $15\ cm^3$
b $85\ cm^3$
c $180\ cm^3$
d $675\ cm^3$
e $5\ m^3$
f $2.6\ m^3$
g $3\ 500\ cm^3$
h $873\ cm^3$

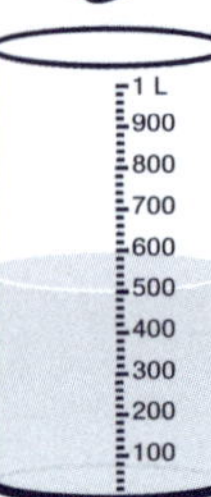

Challenge

A rectangular wading pool measures 1.8 m × 0.8 m and is 45 cm deep. How many litres of water will the pool hold if it is three-quarters full?

Lesson 3 Problem solving

The formula $l \times w \times h$ is used for finding the volume of rectangular prisms. The volume of an object is linked to its capacity.

Help Box

The three-dimensional objects shown here are made up of identical layers and are also called prisms. Each prism is named after the shape of its base. The volume of a prism is calculated by multiplying the height of the prism by the area of its base.

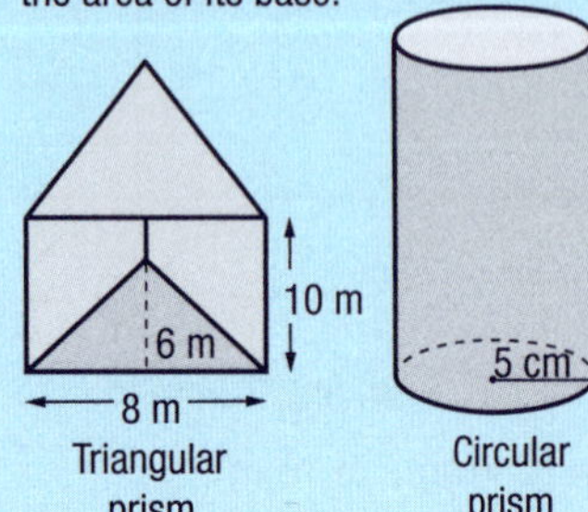

Triangular prism

Circular prism

Triangular prism

Calculate the area of the base

$A = \frac{b \times h}{2}$

$= 8 \times 6 \div 2$

$= 24\ cm^2$

Calculate the volume using the formula

$V = A \times h$

$V = 24 \times 10$

$= 240\ cm^3$

Circular prism

Calculate the area of the base

$A = \pi r^2$

$= 3.14 \times 5 \times 5$

$= 78.5\ cm^2$

Calculate the volume using the formula

$V = A \times h$

$V = 78.5 \times 12$

$= 942\ cm^3$

1 Find the volume of each of the different sized snorkels shown.

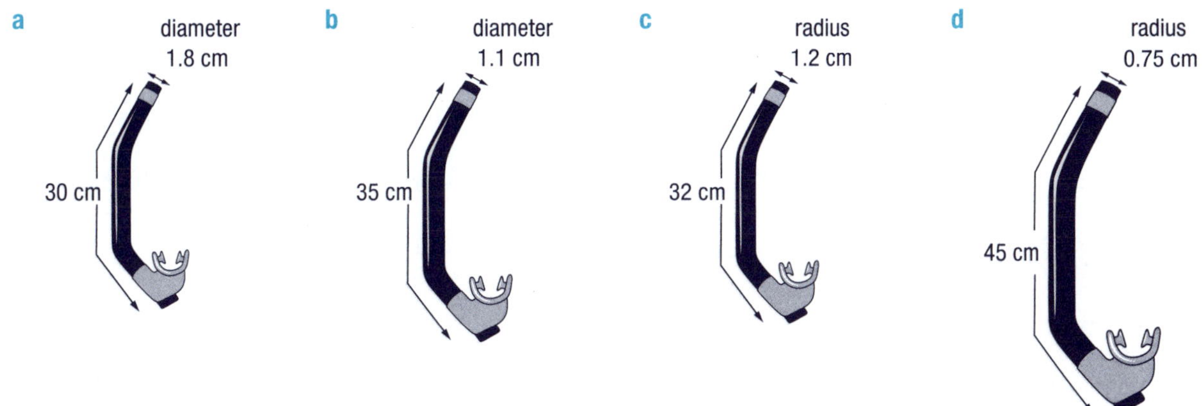

2 Calculate how many litres of outboard motor fuel could be stored in each tank.

a FUEL 0.3 m 0.5 m

b FUEL 40 cm 45 cm

c FUEL 18 cm 10 cm 25 cm

d 5 cm 6 cm FUEL 18 cm 18 cm 18 cm

3 Fish tanks are made by joining together rectangular pieces of glass. Draw and label the dimensions of two different fish tanks that could each hold 20 litres of water in which to keep tropical fish.

4 A rectangular diving pool has dimensions 5.8 m × 5.8 m × 5 m. How many litres of water will it take to fill the pool?

5 The local pool does not have a sloping bottom. It is 25 m long, 12 m wide and 1.5 m deep. How much water will it take to fill the pool to 5 cm below the edge?

6 The cylindrical petrol tank of a motorboat has the capacity to hold 30 litres. What might the dimensions of the tank be? Provide dimensions for two other possibilities.

7 Find the capacity of the following containers.

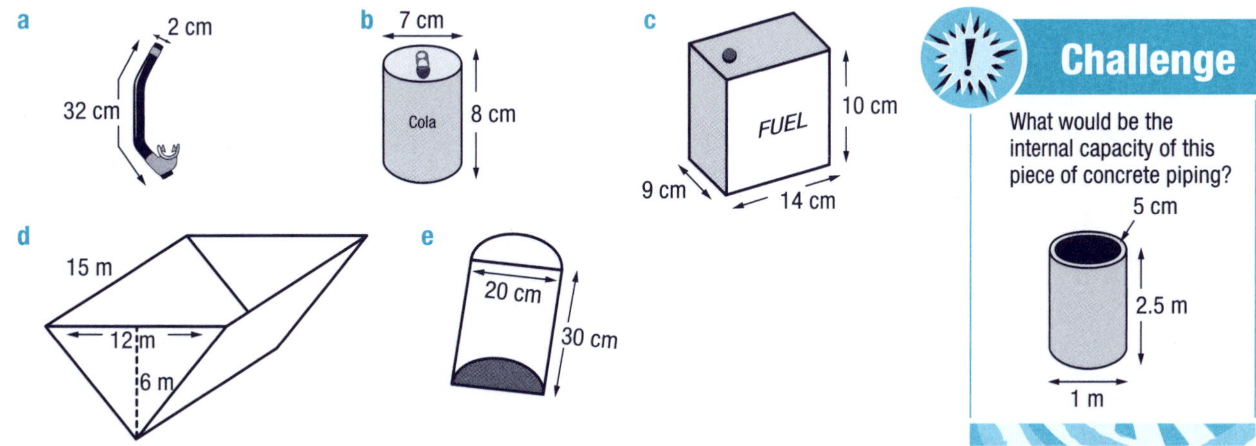

Challenge

What would be the internal capacity of this piece of concrete piping?

Lesson 4 Conversions

The imperial system of measurement uses gallons, quarts, pints and ounces to measure capacity. The metric system of measurement uses litres and millilitres.

The simplicity of the metric system is that it uses multiples of 10 to produce larger or smaller units. For example, we can either say '1000 metres' or 'a kilometre', as kilo means '1000' (which is 10 × 10 × 10). The metric system has become widespread because of the ease of calculating using base 10.

Sometimes it is necessary to convert between imperial and metric measurements. The conversion table shows the measurements rounded to one decimal place.

Measurement of capacity	
Imperial	**Metric**
35 fluid ounces (20 fl. oz = 1 pint)	1 L
44 pints (2 pints = 1 quart)	25 L
1 gallon	4.5 L
44 quarts (4 quarts = 1 gallon)	50 L
11 gallons	50 L

1 Use the table to calculate which is the greater capacity.

a 100 L or 100 gallons
b 100 L or 100 pints
c 100 L or 100 quarts
d 50 L or 50 fl. oz
e 100 mL or 100 fl. oz
f 50 quarts or 50 mL

2 Use the table to convert:

a 500 litres to gallons
b 450 gallons to litres
c 600 litres to quarts
d 440 quarts to litres
e 75 litres to pints
f 132 pints to litres
g 280 fluid ounces to litres
h 25 litres to fluid ounces
i 2 gallons to litres

3
a The fuel tank in a power boat holds 45 litres of fuel. How many gallons is that?
b The icebox has a capacity of 75 litres. How many quarts will it hold?
c The fish tank can hold 66 gallons of water. How many litres will fill the tank?

4 Use your knowledge of volume and capacity to order from most to least:

a 10 L, 10 gallons, 4 quarts, 500 fl. oz
b 10 pints, 1 L, 2 gallons, 100 fl. oz
c 30 fl. oz, 2 gallons, 5 L, 3 quarts
d 5 quarts, 5 L, 100 fl. oz, 4 gallons

5 Four different liquids need to be poured into containers with the dimensions below. Use your knowledge of capacity and volume in metric and imperial measurements to decide which liquid should be placed in each container. Show your working out.

- 1 litre of water
- 10 gallons of petrol
- 8 pints of oil
- 50 litres of milk

a

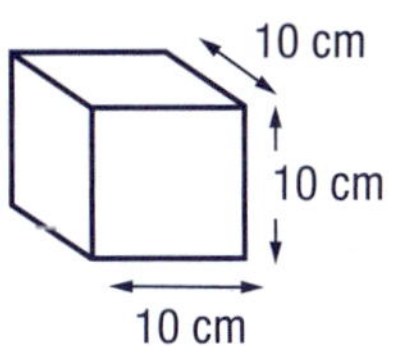

b

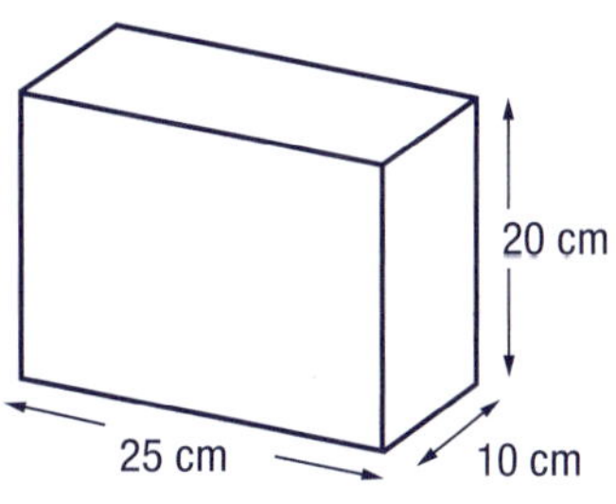

c

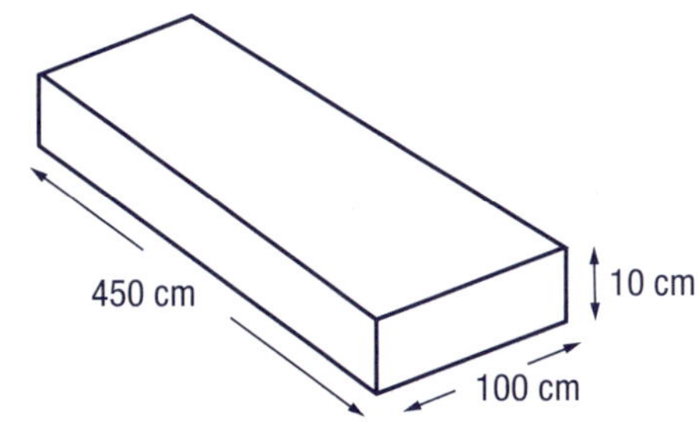

d

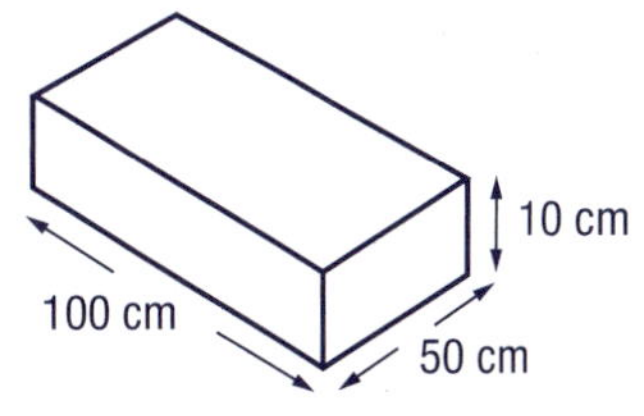

A reasonably accurate conversion between imperial and metric measurements of length can be made using this table.

Measurements of length	
Imperial	**Metric**
1 inch	25 mm
1 foot	0.3 m (30 cm)
1 yard	1 m
1 mile	1.6 km

6 Use the table to calculate which is longer:

a 100 m or 100 feet **b** 10 inches or 100 cm **c** 10 yards or 1 km
d 250 mm or 2 yards **e** 10 miles or 5 km **f** 50 inches or 1 m

7 Use the table to convert:

a 120 cm to feet **b** 20 feet to cm **c** 150 mm to inches
d 40 inches to mm **e** 5 miles to km **f** 32 km to miles
g 150 miles to km **h** 1.2 m to feet **i** 20 feet to m

8 Use your knowledge of metric and imperial measurements to order from longest to shortest:

a 3 mm, 10 inches, 1 yard, 0.3 m **b** 10 m, 0.5 km, 10 yards, 300 cm
c 5 inches, 1 mile, 5 km, 20 yards **d** 15 feet, 5 m, 0.3 km, 50 yards

9 Use the conversion tables in this lesson to complete the following. (Note: The symbol ≈ means 'is approximately the same as'.)

a If 50 L ≈ ____ gallons and 50 L ≈ ___ quarts, then 5 gallons ≈ ____ quarts.
b If 50 L ≈ ____ quarts and 25 L ≈ ___ pints, then 10 quarts ≈ ____ pints.
c If 1 L ≈ ____ pints and 1000 mL ≈ _____ fl. oz, then 10 pints ≈ ____ fl. oz.
d If 25 mm ≈ 1 inch and 1 m ≈ 1 yard, then 10 inches ≈ ____ mm.
e If 3 miles ≈ ____ km and 300 cm = _____ yards, then 1 yard ≈ ____ mm.

Lesson 5 The maths of diving

The popular pastime of scuba diving uses very different skills to those needed to compete in the diving events of competitions such as the South Pacific Games and the Oceania Championships.

The scoring of competition dives requires an interesting four-step mathematical process.

Each diving routine is given a score based on the degree of difficulty.

The degree of difficulty is from

1.2	3.5
(Easy)	(Most difficult)

Each judge gives a score from 1 to 10 for the dive.

Here is the diving score awarded by a panel of seven judges.

Diver: **Harold Selu**
Degree of Difficulty 2.9
Scores 7
7
6.5
7
7
6.5
7
FINAL SCORE: 60.03

Help Box

The four steps involved to calculate the final score are given below.

Step 1: The highest and lowest scores are deleted to remove any possible error by the judges.

Step 2: The mean of the remaining scores is found. This is calculated by adding the scores together and dividing by the number of scores.
$7 + 6.5 + 7 + 7 + 7 = 34.5$
$34.5 \div 5 = 6.9$ The mean score is 6.9.

Step 3: International diving is calculated on the basis of a three-judge panel. The mean score is therefore multiplied by 3 to calculate the score using a three-judge total.
$6.9 \times 3 = 20.7$

Step 4: The three-judge total is multiplied by the degree of difficulty.
$20.7 \times 2.9 = 60.03$ The final score is 60.03.

Diver: **Harold Selu**
Degree of Difficulty 2.9
Scores 7
~~7~~
6.5
7
7
~~6.5~~
7
FINAL SCORE: 60.03

1 Use the four-step process to calculate the final scores for the following divers. Show your working out for each step.

a

Diver: **Wesley Ali**
Degree of Difficulty 2.7
Scores 8
7.5
8
7
8
7.5
7

b

Diver: **Sam Tegeler**
Degree of Difficulty 2.9
Scores 7.5
8
8
8.5
8
8
8.5

c

Diver: **Mino Nalu**
Degree of Difficulty 2.4
Scores 7
7
7.5
7
7
6.5
7

d

Diver: **Joseph Choulai**
Degree of Difficulty 2.5
Scores 7.5
8
9
8
8.5
8.5
8.5

e

Diver: **Aaron Cockrem**
Degree of Difficulty 2.8
Scores 8
8
8
8
8
8
8

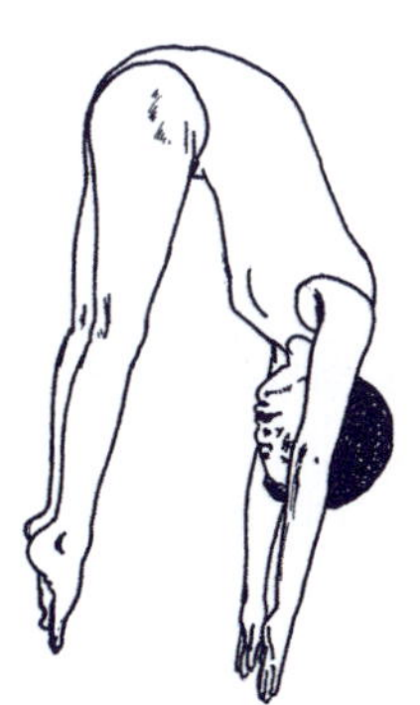

2 Calculate what Joseph's final score would have been if his dive's degree of difficulty was 2.9 instead of 2.5.

3 Calculate what Sam's final score would have been if his dive's degree of difficulty was 2.4 instead of 2.9.

4 How different would Aaron's final score have been if one of the judges had scored him at 6 instead of 8?

5 How different would Mino's final score have been if the judge who scored him at 6.5 scored him at 8 instead?

6 **a** A diver has decided to do a dive with a degree of difficulty of 2.6. The scores from five judges so far are 7, 8, 7.5, 8, 8. Suggest two other scores that would enable the diver to gain a final score of at least 60.

b Calculate the final score for the diver using the numbers you suggested in part a as two of the seven judges' scores.

Challenge

Calculate a set of seven judges' scores that would enable a diver to gain a final score of between 68.5 and 71 when doing a dive with a degree of difficulty of 2.7.

7 Remembering that international diving is calculated on the basis of a three-judge panel, calculate the final score for these dives that used a panel of six judges (rather than seven as in the previous examples).

a

Diver: **A**
Degree of Difficulty 2.3
Scores 8.5
8
9
8
8.5
9

b

Diver: **B**
Degree of Difficulty 2.7
Scores 7.5
7.5
7
7.5
7.5
7

c

Diver: **C**
Degree of Difficulty 2.1
Scores 9
9
9
8.5
8.5
9

8 Some spectators were trying to decide if it would be better to get all scores of 10 for a dive with a low degree of difficulty (1.2) or all scores of 7 for a dive with a high degree of difficulty (3).

a Which option do you think would be best?

b Calculate the final score for each option.

Lesson 6 Swimming sensations

Papua New Guinea has produced some world-class swimmers, including Ryan Pini who has won gold, silver and bronze medals in the Olympic and Commonwealth Games and the Oceania Championships. During this lesson you will investigate whether you can walk as fast as champions can swim and how swim-times vary across events.

In a small group, measure and mark out a 10-metre length in the yard. Use a stopwatch to time each other to walk 10 metres. Times should be to two decimal places if possible. Each student should walk the distance three times at a normal walking pace and then calculate the mean of their times to find their average walking pace over 10 metres.

1 Copy and complete the table as a group.

Student's Name	Time taken Walk 1	Time taken Walk 2	Time taken Walk 3	Mean time for 3 walks

Pini broke the men's 50 m backstroke record at the 2004 Oceania Championships with a time of 26.83 seconds.

2 What speed does this work out to per second?

3 Assuming his speed was constant over the distance, how long did Pini take to swim 10 m?

4 Write a sentence to describe the difference between your walking pace and Pini's backstroke pace over 10 m.

The US women's 4 × 200 m freestyle relay team also broke a record at the Athens Olympics. The four women won gold in 7 min 53.42 s (7:53.42). The previous record of 7:55.47 was set by East Germany in 1987.

5 By how much did the US team beat the record?

6 What was the mean time each woman in the US team took to swim her 200 m?

7 Calculate your walk-time over 200 m and write a sentence to compare it to that of the US team's mean swim-time.

8 One team member, Natalie Coughlin, swam her section in a time of 1:57.74. How much faster was she than the team's mean swim-time?

9 Now that you know Coughlin's swim-time, calculate the mean swim-time of each of the remaining members of the team.

10 If the four members of the team had been able to swim at the same pace as Coughlin, how long would they have taken to complete the race?

11 The Chinese team came second in the race with a time of 7:55.97. How much slower were they than the US team?

In the 2006 Oceania Championships Jane Mopio and Nicolle Ellesworth both represented Papua New Guinea in the women's 100 m freestyle trials. Mopio scored a time of 1:00.72 and Ellesworth swam the distance in 1:05.44. The winner of the heat swam 100 m in 59.91.

12 What was the time difference between Mopio's and Ellesworth's swims?

13 If Mopio swam at a constant speed for the 100 m, how long did she take to swim 10 m?

14 If the winner of the trial swam at a constant speed for the 100 m, estimate how far she had swum after 10 seconds. Explain your thinking.

15 Write a sentence to describe the difference between your walking pace and Mopia's freestyle pace over 10 m.

16 The winner of the heat swam 100 m in a time of 59.91. How much did the winner beat Ellesworth by?

17 Some public swimming pools are 25 m long, while others are 50 m long. Calculate how long it would take you to walk around the outside edge of a pool at your normal walking pace if it was:

- **a** 50 m long and 30 m wide
- **b** 25 m long and 15 m wide
- **c** 50 m long and 25 m wide
- **d** 25 m long and 20 m wide.

Challenge

How far could Pini swim if he was swimming at a constant speed of 1.95 metres per second while you walked each of the distances described in question 17?

Lesson 7 Percentage change

1 A village has a population of 900 people. Twenty-eight per cent are adult males and 24% are adult females. How many people in the village are not adults?

2 Solve the following problems.

- **a** In a school 18% of 450 students can swim. How many students cannot swim?
- **b** A fuel tank with a capacity of 55 L is 35% full. How many litres are in the tank?
- **c** A sports stadium can hold 50 000 people when full. How many people are in the stadium if it is 72% full?

Help Box

At times, a percentage change in data can be reported.

For example: 'The swimmer improved her time in the event by 15% between trials 1 and 2'.

To increase a number by 15% we multiply it by (100 + 15)%, that is, 115% (or 1.15).

To decrease a number by 15% we multiply it by (100 – 15)%, that is, 85% (or 0.85).

So, to increase 59 by 15% ⟶ $59 \times 1.15 = 67.85$
new value is 67.85

To decrease 59 by 15% ⟶ $59 \times 0.85 = 50.15$
new value is 50.15

3 What would you multiply by to increase a number by:

a	10%	**b**	15%	**c**	38%
d	7%	**e**	4.7%	**f**	17.5%?

4 What would you multiply by to decrease a number by:

a	25%	**b**	18%	**c**	46%
d	3%	**e**	2.8%	**f**	11.8%?

5 Increase:

a	38 by 25%	**b**	57 by 36%
c	82 by 3%	**d**	112 by 17%

6 Decrease:

a	45 by 15%	**b**	32 by 18%
c	73 by 6%	**d**	102 by 31%

7 A swimming coach wants the team to increase the number of laps they swim each training session by 15%. How many laps will each swimmer now need to swim per session?

- **a** Betty currently swims 20 laps
- **b** Samson currently swims 25 laps
- **c** Tom currently swims 45 laps
- **d** Ivy currently swims 32 laps

Help Box

The number of students at a school who can swim has increased from 28 to 35. What is the percentage increase?

$$\text{Percentage increase} = \frac{\text{increase}}{\text{original number}} \times 100$$

So the percentage increase of children who can swim is $\frac{7}{28} \times 100 = 25\%$.

8 This table shows the population of a town.

Year	Men	Women	Children
2004	500	400	200
2005	600	450	240
2006	700	500	300
2007	720	560	320

- **a** What was the percentage increase of men from 2004 to 2005?
- **b** What was the percentage increase of children from 2005 to 2006?
- **c** What percentage of the population in 2007 were women?

9 Calculate the percentage increase when:

- **a** 60 becomes 75
- **b** 40 becomes 48
- **c** 9 becomes 27
- **d** 35 becomes 140
- **e** 36 becomes 45
- **f** 140 becomes 196

Challenge

Create a similar table for the village that began with the same data for 2004 and then showed a continual growth of 15% per annum across all three sectors for the next three years.

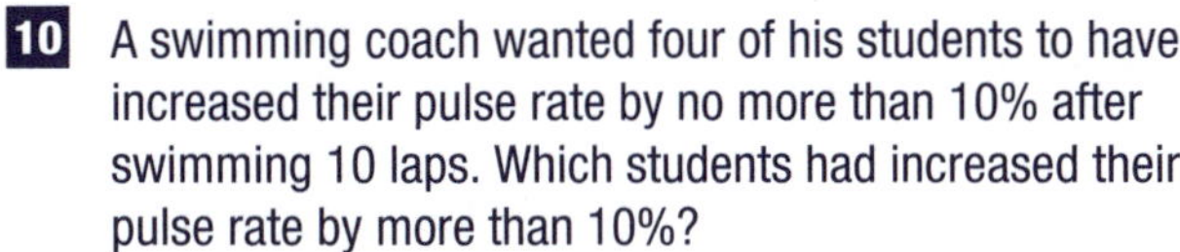

10 A swimming coach wanted four of his students to have increased their pulse rate by no more than 10% after swimming 10 laps. Which students had increased their pulse rate by more than 10%?

- **a** Fabian was 22 beats per 10 seconds, now 24.
- **b** Ryan was 20 beats per 10 seconds, now 23.
- **c** Wesley was 23 beats per 10 seconds, now 25.
- **d** Joseph was 20 beats per 10 seconds, now 24.

Help Box

The number of students in the swimming team fell from 20 to 16. What is the percentage decrease?

$$\text{Percentage decrease} = \frac{\text{decrease}}{\text{original number}} \times 100$$

So the percentage decrease in students is $\frac{4}{20} \times 100 = 20\%$

11 Calculate the percentage decrease when:

- **a** 36 becomes 27
- **b** 75 becomes 60
- **c** 25 becomes 19
- **d** 400 becomes 300
- **e** 56 becomes 42
- **f** 110 becomes 99

12 A sports store was having a 'minimum 15% off everything' sale and had marked some items with the following price tags. Check that the owner had reduced the items correctly.

- **a** a swimsuit was K50, sale price K42
- **b** a sports bag was K20, now K17
- **c** a snorkel was K12, now K10
- **d** a pair of flippers was K18, now K16

Lesson 8 Statistical average

Averages are commonly used in sporting events to measure performance. In Grade 7 you learned to calculate the mean of a set of data. The mean (often called the average) is the most commonly used measure in statistics.

Help Box

The **mean** of a set of scores is found by dividing the sum of the scores by the number of scores.

$\text{mean} = \frac{\text{sum of scores}}{\text{number of scores}}$

So the mean of (2, 3, 7) is $\frac{12}{3}$
$= 4$

1 Find the mean of the following sets of data to one decimal place.

- **a** 2, 3, 5, 6
- **b** 23, 15, 19, 7
- **c** 1.8, 5, 8, 9.2
- **d** 2.4, 3.1, 4, 1.2

2 Find the missing numbers in the following data sets if each set has a mean of 5.

- **a** 7, ___
- **b** 1, 3, 8, ____
- **c** 3, 6, 7, 9, _____
- **d** 1, 2, 16, ___

3 A set of four numbers has a mean of 18. What might the four numbers be?

4 Below are the time trials for two swimmers competing in the 50 m backstroke. Calculate the mean swim time for each swimmer.

Swimmer A		Swimmer B	
	28.42 s		27.83 s
	29.04 s		28.26 s
	27.94 s		29.13 s
	29.08 s		28.74 s

Another way of measuring the middle of a set of data is to calculate the median.

Help Box

The **median** of a set of scores is the middle score when the scores are placed in order from smallest to largest. If the number of scores is even, the median is the average of the two numbers in the middle.

The median of the data set (2, 5, 14) is 5.

The median of the data set (2, 3, 6, 11) is 4.5.

5 Find the median of the following sets of data.

- **a** 12, 6, 11, 2, 5
- **b** 15, 4, 19, 6
- **c** 5, 2, 8, 9, 15, 7, 8
- **d** 2, 4, 3, 5, 4, 8
- **e** 7, 11, 18, 3, 23, 6, 4, 19, 12
- **f** 2, 5, 21, 4, 3, 15, 8, 16
- **g** 1.2, 4.5, 2.06, 7.03, 2.6

6 A set of four decimal numbers has a median of 3.5. What might the four numbers be?

7 Find the mode of the following data sets.

a 2, 6, 9, 2, 3
b 11, 4, 56, 4, 11
c 15, 2, 4, 11, 15, 8
d 1, 3, 3, 7, 4
e 7, 5, 19, 3, 4
f 2, 5, 2

8 a Create a data set of six numbers that has 7 as the mode.

b Create a data set of six numbers that has two modes.

c Create a data set of six numbers that does not have a mode.

9 Copy the following table into your workbook and complete by finding the mean, median and mode for each set of data.

	Data set	Mean	Median	Mode
a	23, 4, 7, 4, 9			
b	2, 5, 15, 8, 11, 12			
c	1.3, 3.2, 1.03, 4, 1.3			
d	10.3, 9.2, 9.01, 10.03			
e	11, 15, 11, 23, 9, 7, 9			

Help Box

The **mode** is the score that appears most frequently in a set of scores. There can be more than one mode in a set. If each score appears only once there is no mode.

The mode of the data set (2, 4, 4, 5, 7) is 4.

The mode of the data set (3, 3, 5, 6, 6, 7, 9) is 3 and 6.

There is no mode in the data set (2, 4, 5, 8).

10 A diver gained these scores during a diving competition: 70.5, 63.5, 72, 73.5, 62, 72, 73.

a What was her mean score for the competition?

b What was her median score?

c What was her mode score?

d Which of the mean, median and mode best represents her dives? Explain your answer.

Lesson 9 Data displays

Data can be presented in various ways. It is as important to be able to present data in tables and graphs as it is to be able to interpret data.

Method of presenting data	Bar & Column graph	Pictogram	Line graph	Pie chart	Table
Description	column or bar length represents a specific amount	uses pictures to create columns	shows continuous data, like temperature	each category is shown as a fraction of the circle	list data strategically
Example	Numbers / Years; Height / Time	Height; 🌳 = 10 m	Temperature / Time	A, B, C, D	Student / Amount: Bill 24, John 16, Lilly 13, Mary 6

1 a Draw a pictogram to display the total number of students competing in each event at the national swimming championships. The data is given in the table on the right.

b Write three statements that compare some of the data in the graph.

	Event	No. of competitors
1	50 m freestyle	32
2	100 m freestyle	46
3	50 m butterfly	14
4	50 m backstroke	36
5	800 m freestyle	2
6	200 m freestyle	22
7	4 × 100 relay	28
8	4 × 200 relay	20

2 The table below shows the pulse rate of a swimmer during the 100 m freestyle event.

a Draw a line graph to display the data.

b Describe what happens to the swimmer's pulse rate during the race.

c Predict what the swimmer's pulse rate might be at the 100 m mark.

Distance from start (metres)	15	30	45	60	75	90
Pulse (p/min)	80	120	120	135	150	160

3 This table shows the number of medals awarded to children in the Aqua Swim Club.

	Name	Medals
1	Patricia	6
2	Rosa	1
3	Marlon	4
4	Peter	2
5	Mary	3
6	Harold	4

- **a** Display the data in a pie chart.
- **b** What fraction of the chart is each student's medal tally?

4 This column graph shows the diving scores achieved by Rebecca during trials.

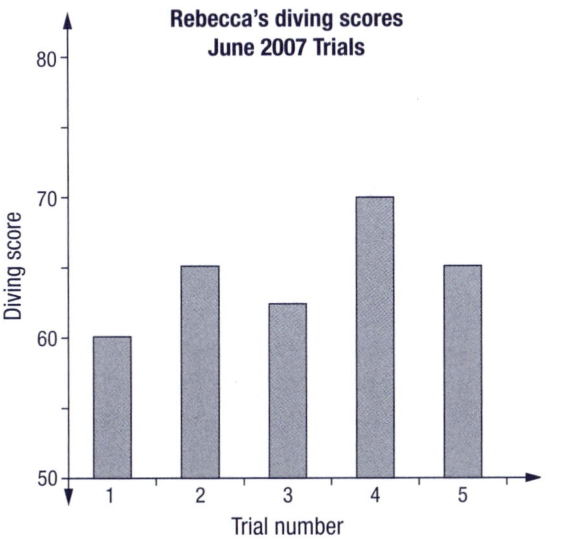

- **a** What does the vertical axis show?
- **b** What does the horizontal axis show?
- **c** Which dive was Rebecca's best?
- **d** What was the difference between her best and worst scores?
- **e** Predict the range you think Rebecca's next trial score is likely to be within.

Peter went snorkelling for 30 minutes. For the first 5 minutes he only saw coral. Suddenly, a huge marlin appeared and gave him a fright. The marlin moved on after 3 minutes leaving Peter to watch a school of small tropical fish. As he was returning to the boat his favourite fish, a clownfish, swam out from between the rocks. Peter stayed to watch it for a while.

5 Alice drew graph A and Rodney drew graph B to describe the changes in Peter's pulse rate while he was snorkelling.

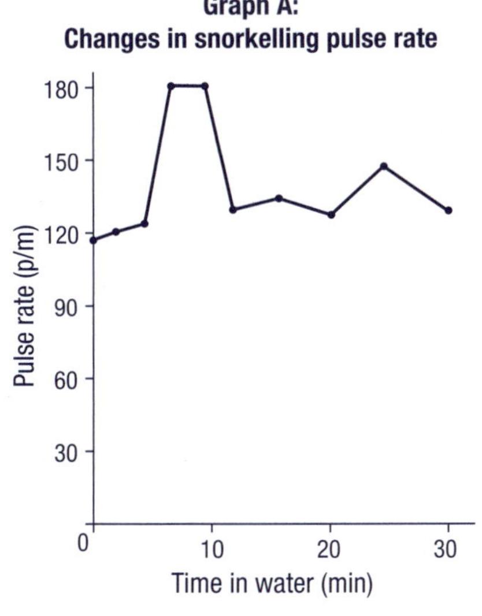

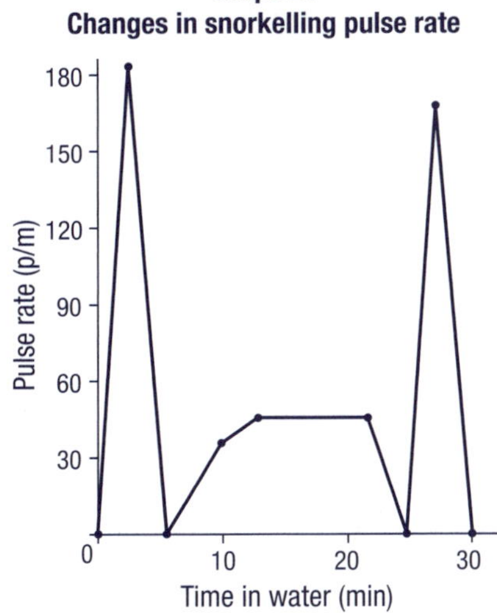

- **a** What is shown on the horizontal axis?
- **b** Both graphs begin at different places on the *y*-axis. Discuss with a friend whether graph A or graph B is more likely to begin in the correct place. Explain what you decided in writing.
- **c** Which of the two graphs shows the greater range in the pulse rate? Explain how you decided.
- **d** Which graph do you think is probably a good representation of what Peter's pulse rate would have been for the 30 minutes described?
- **e** What changes would there be to the graph if Peter hadn't seen the marlin?

6 a Write a title for the graph on the right and label the two axes appropriately.

b Write a few sentences to describe the information in the graph.

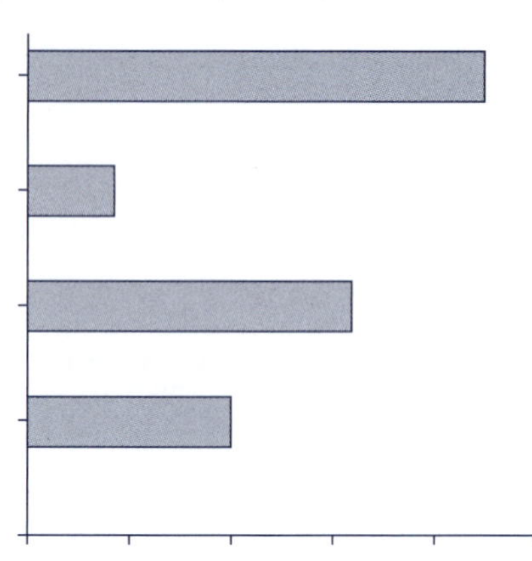

7 a The graph on the right shows the distance that Rachel was from Dara while the two girls were snorkelling. How far away was Dara when they began snorkelling?

b How far apart were the two girls when the line on the graph first flattens out?

c During what time interval were the girls furthest apart?

d How long were the girls snorkelling?

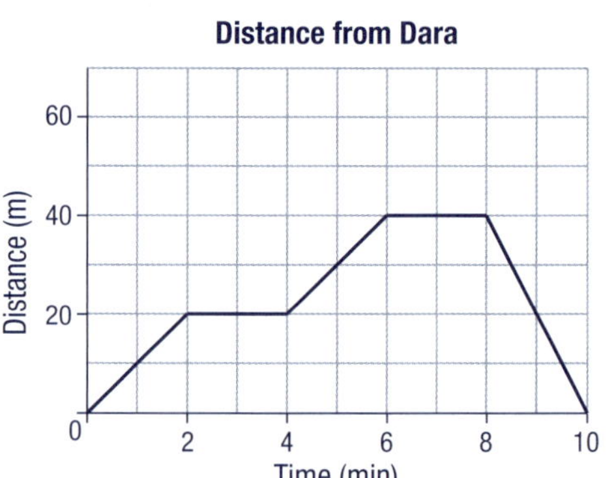

8 This graph shows Darnley's swim times across eight attempts at the 200 m freestyle. Convert the information displayed in the graph into a table.

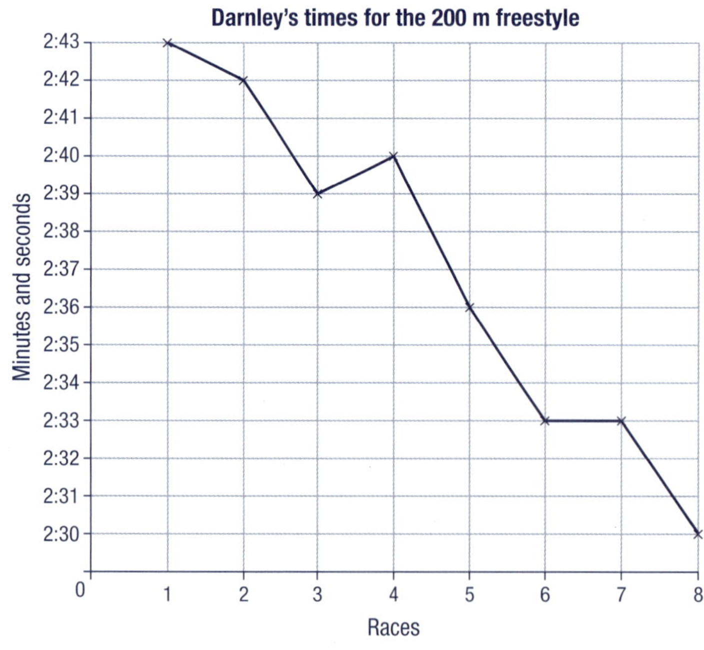

Lesson 10 Misuse of statistical data

Sometimes the scales used on a graph can give a false impression about what the data actually shows. It is important to examine the graph carefully in case it has been accidentally or deliberately drawn to give the wrong impression.

1 a The following two graphs were drawn using the same data. What differences do you notice in the axes of the two graphs?

b At first glance at graph B, what might you think about Sam's medal tally?

c Which graph do you think presents the data most appropriately? Why?

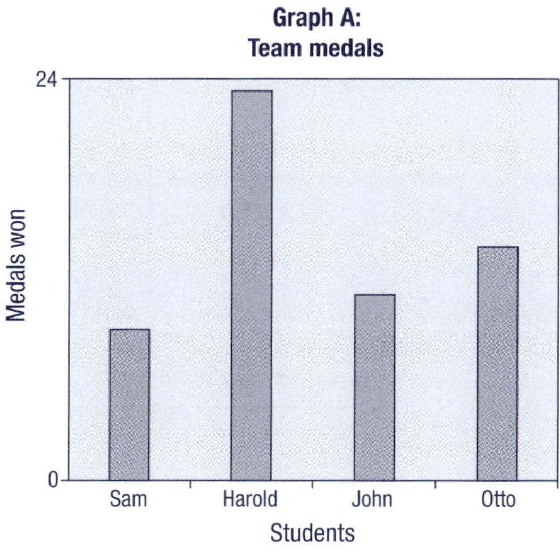

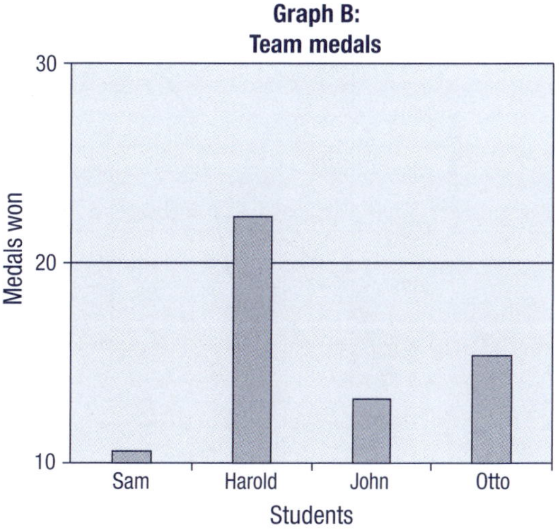

2 The following two graphs show the number of people who went fishing with Acme Boat Charter Company over a six month period.

a

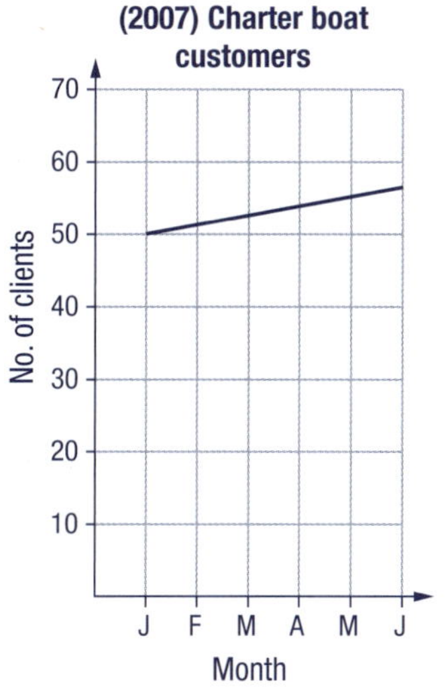

b

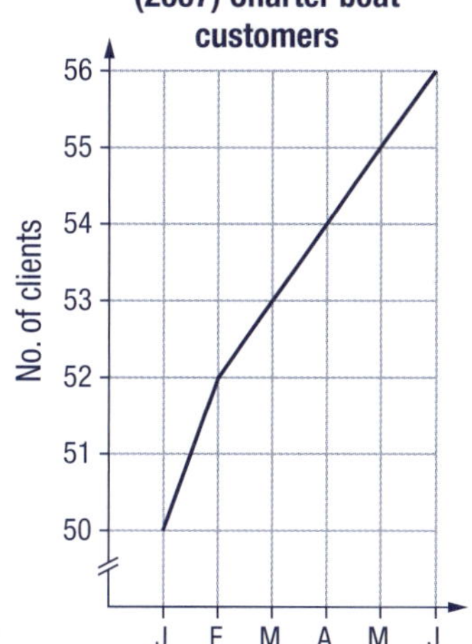

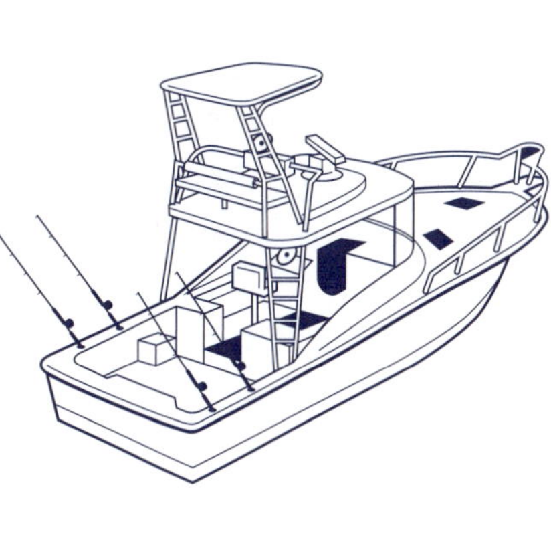

- **a** Do both graphs show the same information?
- **b** How many people went fishing with Acme in March?
- **c** Which month did Acme have the most customers?
- **d** What impression does Graph B give compared to Graph A?
- **e** Explain how the *y*-axis of Graph B has been used to create a misleading impression.
- **f** Suggest a situation when Acme might deliberately use Graph B to create a false impression.

3 The following line graphs were drawn by two different people: one who feels that there are too many diving events during the swimming championships and another who enjoys watching the diving immensely.

a

Graph A: Spectators at diving events

No. of people: 50, 100, 150, 200, 250, 300, 350

Survey time: 8am, 9am, 10am, 11am, 12pm, 1pm, 2pm, 3pm

b

Graph B: Spectators at diving events

No. of people: 200, 400, 600, 800, 1000, 1200, 1400

Survey time: 8am, 9am, 10am, 11am, 12pm, 1pm, 2pm, 3pm

- **a** Do both graphs show the same information?
- **b** Which of the two people do you think drew graph A? Why?
- **c** What makes the graphs appear so different?
- **d** According to both graphs, at what time of day does the diving have the most spectators?
- **e** According to both graphs, how many spectators are at the diving at 9am?

4 The graph opposite shows how popular certain swimmers are according to a phone survey.

- **a** How many more votes did swimmer B get than swimmer D?
- **b** By how many votes did swimmer C beat swimmer A?
- **c** How does the graph mislead people to think that swimmer B more than doubled swimmer A's votes?
- **d** In what situation might someone deliberately choose to create a graph like this?
- **e** Redraw the graph with the *y*-axis starting from zero and comment on the different impression the graph now creates.

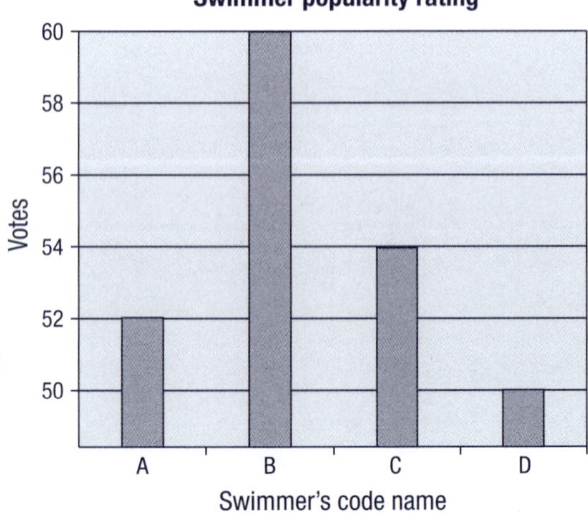

5 This graph shows the number of medals a club won for three different types of events at the local aquatic carnival.

a What was the total number of medals that the club won across the three events?

b How is the pictogram misleading?

c Redraw the pictogram so that it is not so misleading.

6 Find a graph in the local newspaper or ask a friend to draw a graph. Paste it into your workbook. Redraw the graph to provide the same information but in a way that may be misleading.

Learning Unit 3 In the Bag

Strand: Number and Application

Fractions	Outcome 8.1.1	Apply fractions in problem solving

Strand: Space and Shape

Volume	Outcome 8.2.6	Investigate volumes of cylinders, cones and pyramids and apply some volume rules
Nets	Outcome 8.2.13	Associate nets with the solids they form

Strand: Patterns and Algebra

Packing	Outcome 8.5.1	Apply packing patterns in solving problems from real life

Lesson 1: Introduction	Investigating packaging ideas
Lesson 2: The shape of things	Investigating the features of prisms, pyramids, cylinders and cones
Lesson 3: Opening it out	Drawing nets of solids Constructing solids from nets Identifying nets
Lesson 4: The space inside	Finding the volume of rectangular prisms and cylinders Calculating the volume of uniform solids Linking capacity and volume
Lesson 5: Tapering solids	Calculating the volume of cones and pyramids
Lesson 6: Problem solving with volume	Working with volume and capacity Applying formulae to find volume
Lesson 7: Packing efficiently	Stacking items efficiently Using area to make seating plans
Lesson 8: Bits and pieces	Resizing using fractions Solving problems with fractions Using the four processes
Lesson 9: Fractions and decimals	Solving problems

Lesson 1 Introduction

In this unit you will look at a range of sporting equipment and solve problems relating to the packaging and stacking of different shaped items. You will extend your knowledge of capacity and volume, and construct nets of various three-dimensional solids.

Manufacturers look at the shape of an item when trying to decide what packaging they should use. The packaging needs to protect the item and be a good enough fit so that there is minimal wastage of material when making the container. Packaging is usually shaped to enable the item to be stacked efficiently for transportation and on the shelves in shops. This photo shows some of the packaging found in supermarkets.

1 Tennis balls are often packed and sold as three balls in one tube.

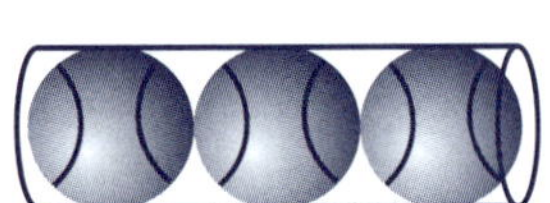

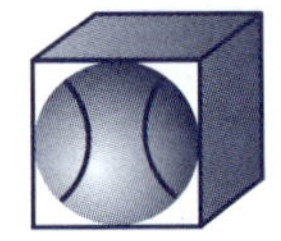

- a Discuss with a friend whether this is more or less efficient than packing each ball individually into a small box of its own.
- b Experiment with three balls to test your ideas.
- c Write a few sentences to describe what you found.

2 Shopping bags come in various shapes and sizes.

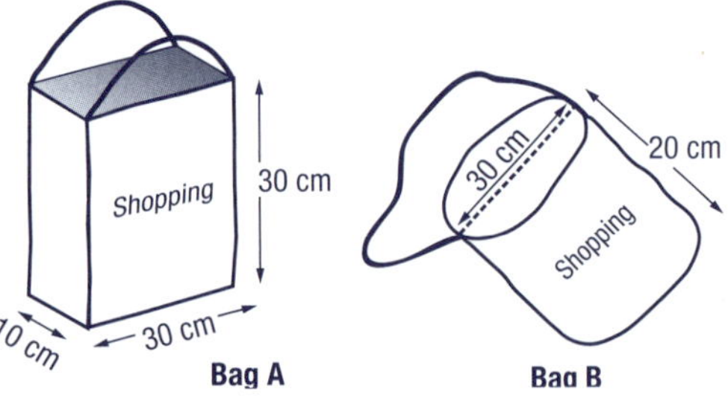

- a Investigate and discuss with a friend which of the two shopping bags shown here would be most practical if you were going to buy six large bottles of cola from the supermarket.
- b Write a few sentences to explain your decision.

3 Some sports drinks are sold in bottles and some are sold in cans.

- a Which container would be more practical for stacking on shelves? Why?
- b What advantages might there be to the manufacturer for packaging drinks in the other container?

Lesson 2 The shape of things

Three-dimensional shapes are known as **solids**. This is a collection of solids.

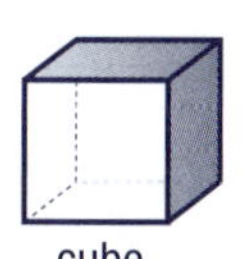
cube

cone

sphere

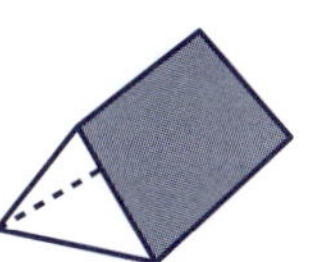
triangular prism

cylinder

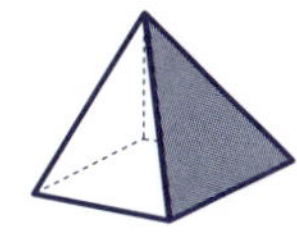
square-based pyramid

Some solids have plane surfaces. (A 'plane' surface is a mathematical word meaning a 'flat' surface.) Other solids have curved surfaces, while some solids are a combination of both plane and curved surfaces.

Help Box

A **cube** is a solid with only plane surfaces.

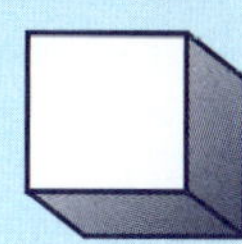

A **sphere** is a solid with only curved surfaces.

A **cylinder** is a solid with both curved and plane surfaces.

1 Here is a collection of different solids and a table. Copy the table into your workbook. Complete it by listing the letter beside each shape in the appropriate column.

Solids with:		
all plane surfaces	all curved surfaces	a combination of plane and curved surfaces

A 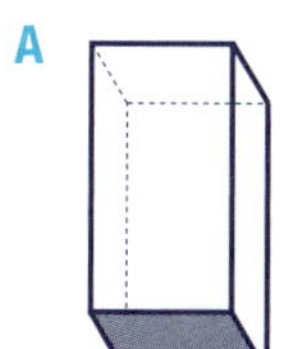

B 

C 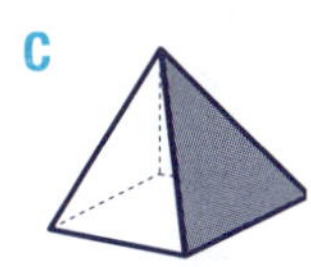

D 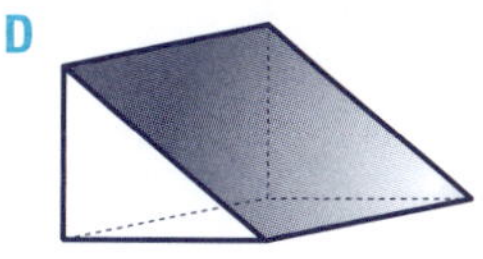

E 

F 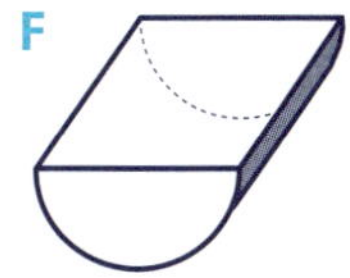

G 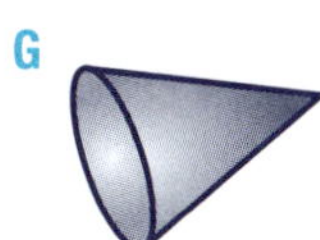

H 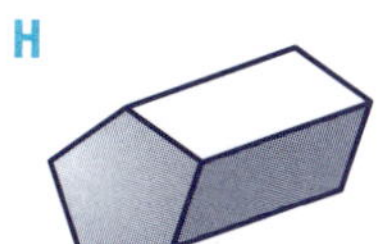

I 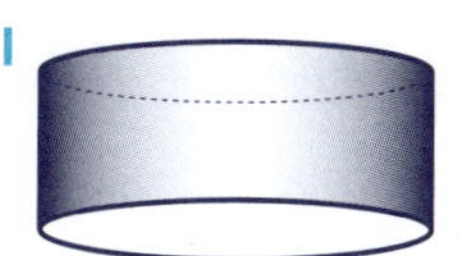

J 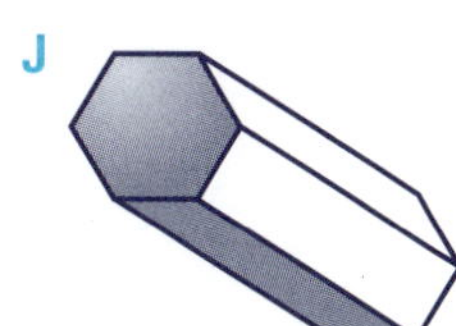

K 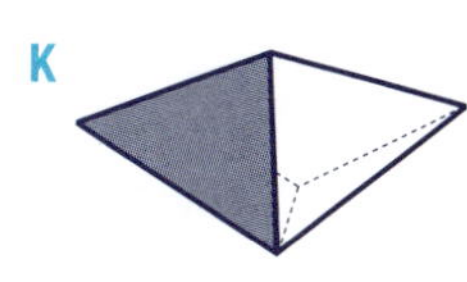

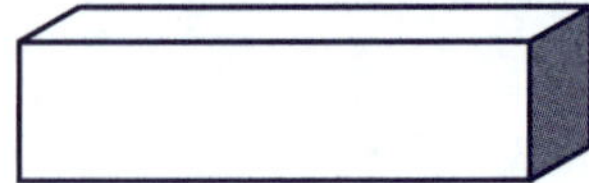

A **prism** is a solid that has two matching polygons on its ends. All other faces in a prism are plane, rectangular shapes. Prisms are named according to the shape of their ends. A rectangular prism is one example of a prism.

2 a Draw three different prisms that are not rectangular prisms.

b Label each prism with its correct name.

c List the names of the two-dimensional faces of each prism that you drew.

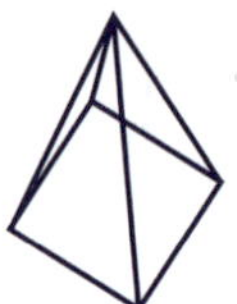

A **pyramid** is a solid with a polygon as a base. All other sides in a pyramid are flat, triangular shapes that meet at an apex. Pyramids are named according to the shape of their base. Here is a square pyramid.

3 a Draw three different pyramids that are not square pyramids.

b Label each pyramid with its correct name.

c List the names of the two-dimensional faces of each pyramid that you drew.

The curved surfaces of a cone and a cylinder mean they cannot be classified as either a prism or a pyramid.

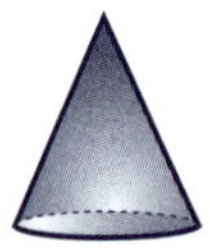

Challenge

Draw a solid that is not a prism, pyramid, cylinder or cone.

Write a few sentences to explain why your solid cannot be classified as one of the solids listed above.

4 List the geometrical name and the number of two-dimensional shapes that are needed to make each of the following solids:

a triangular pyramid b cube c cylinder d rectangular prism

5 Record the geometrical name that you would use to describe the shape of the following:

a

a can of drink

b

a house

c

a shoebox

d

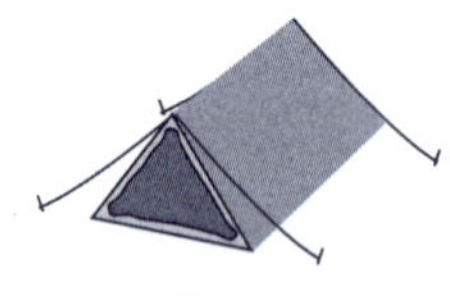

a tent

e

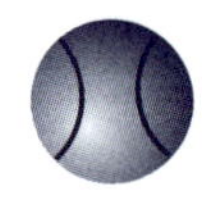

a ball

f

a party hat

g

a pipe

h

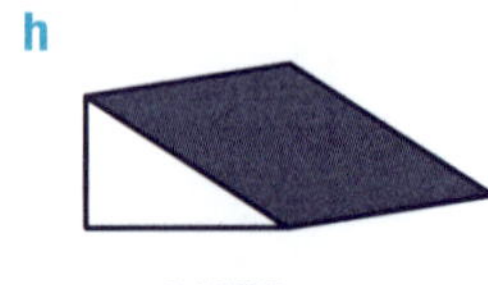

a ramp

6 Give an example of an item whose shape could be described as a:

a triangular pyramid b rectangular pyramid c sphere

d triangular prism e rectangular prism f cone

Lesson 3 Opening it out

A **net** is a two-dimensional shape that can be cut out and folded to make a solid. To the right are two different nets that can be folded to form a rectangular prism. One is drawn on grid paper and one was drawn using a ruler and a protractor to make 90° angles in each corner.

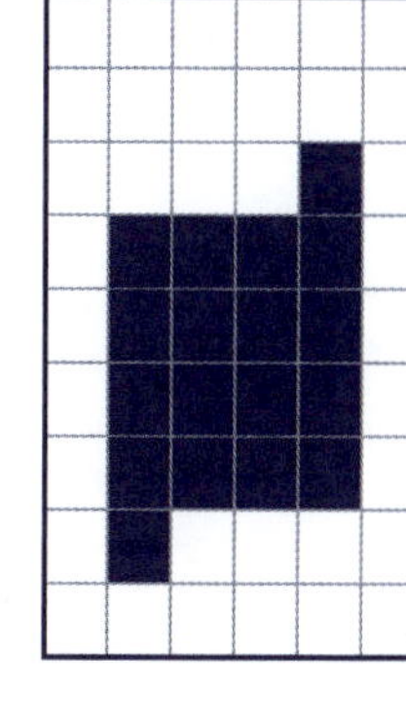

1 a Use a ruler and protractor to draw a net for a rectangular prism that is different from those shown here. Cut out the net and fold it to make the rectangular prism.

b List the geometrical name of each of the faces on the prism.

c Does this solid stack easily? If so, illustrate with a diagram to show how a collection of rectangular prisms could be stacked together efficiently.

Like the rectangular prism, some other solids also have more than one possible net. Here are two nets for another solid.

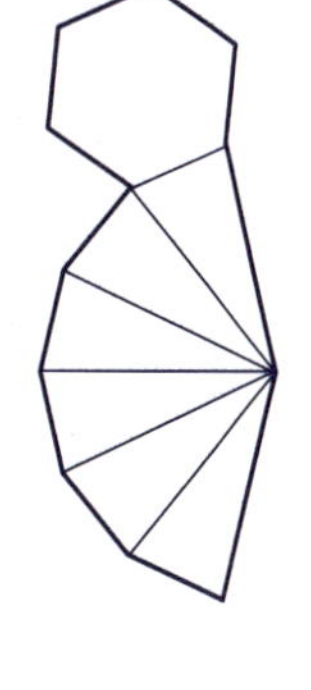

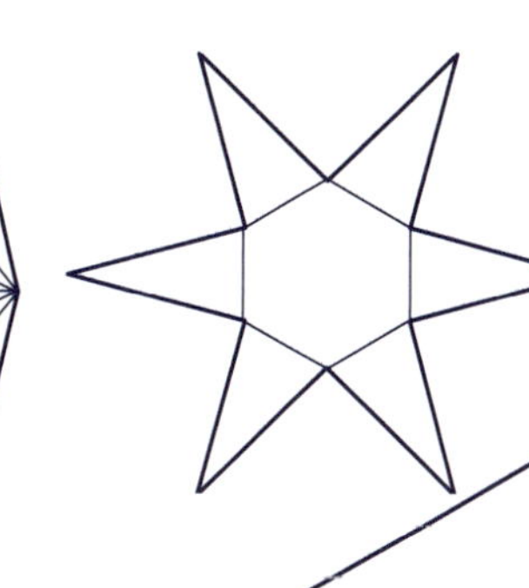

2 Trace the shapes below onto cardboard.

a Use the two shapes together to make a template that could be used to make a net for a hexagonal prism.

b Use the two shapes together to make a template that could be used to make three other nets for hexagonal prisms different to the net you drew for the previous question.

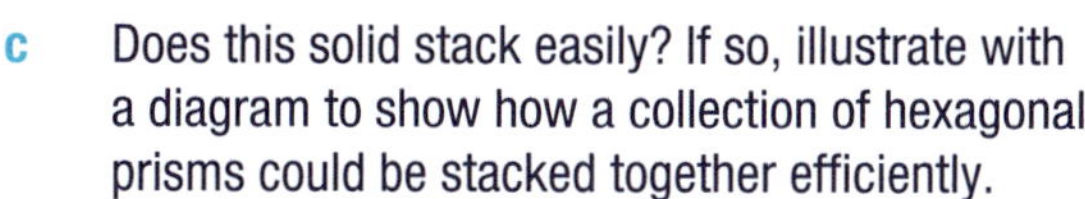

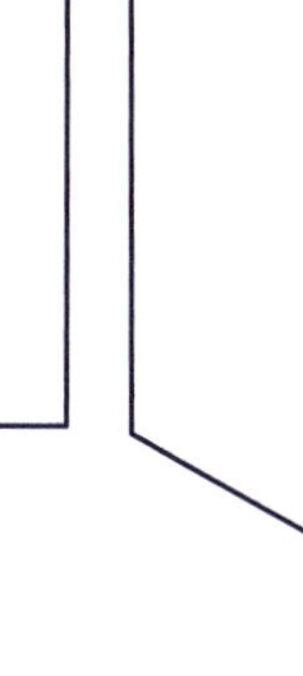

c Does this solid stack easily? If so, illustrate with a diagram to show how a collection of hexagonal prisms could be stacked together efficiently.

3 **a** Trace this net onto paper and cut it out. Fold it into a solid.

b Name the solid.

c Does this solid stack easily? If so, illustrate with a diagram to show how a collection of these could be stacked together efficiently.

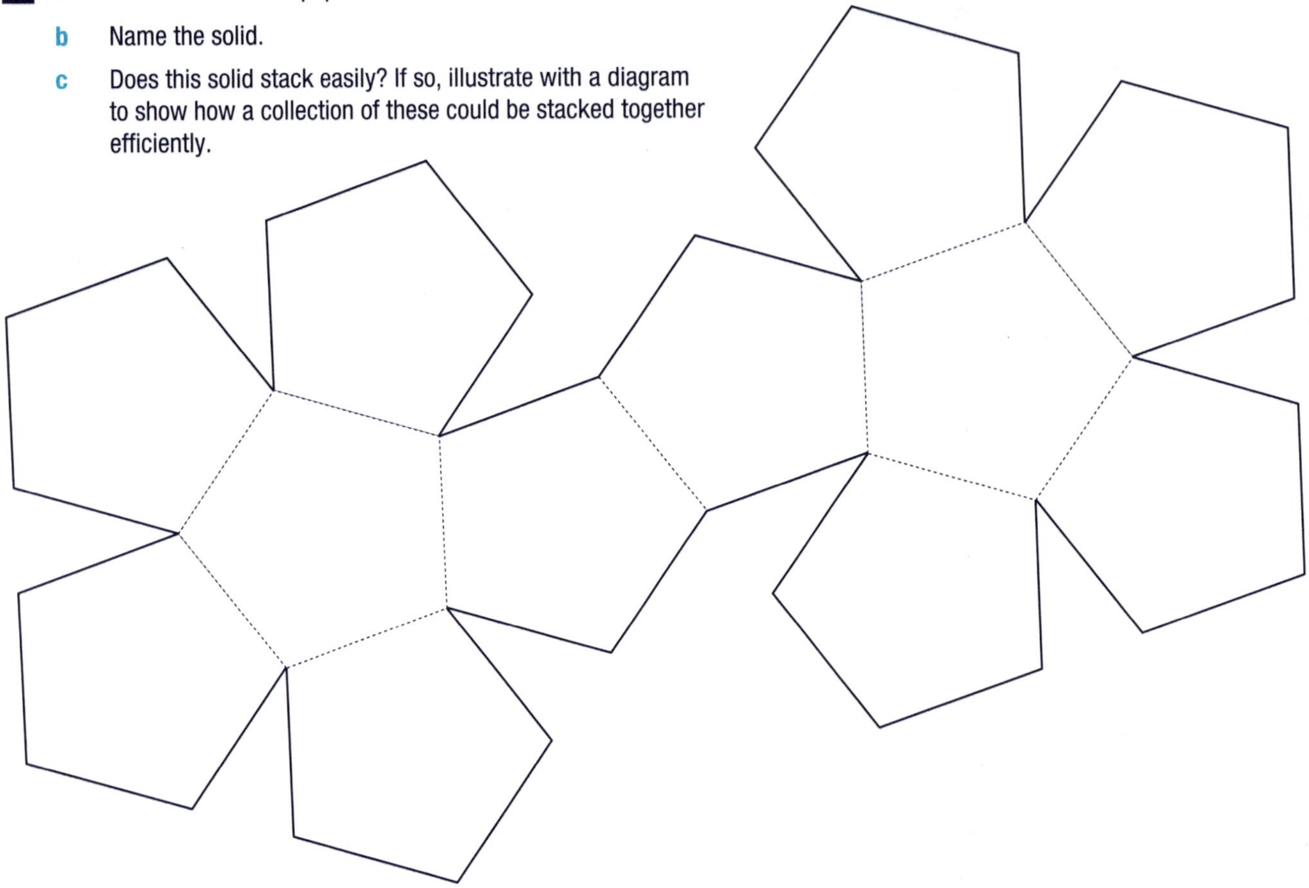

4 Draw a net for each of the following solids.

a

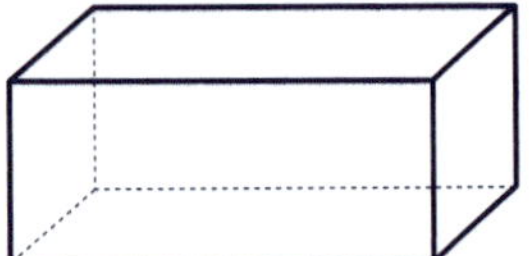

b

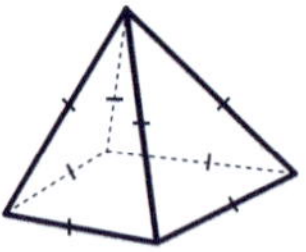

c

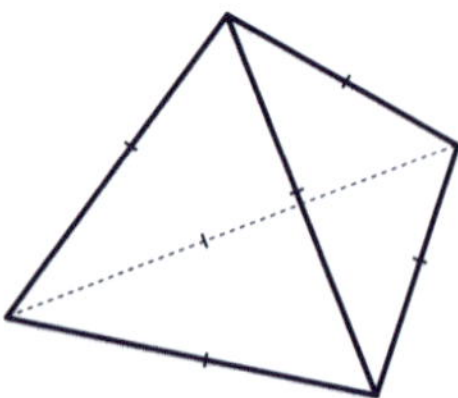

d

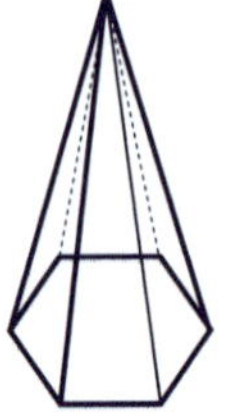

e

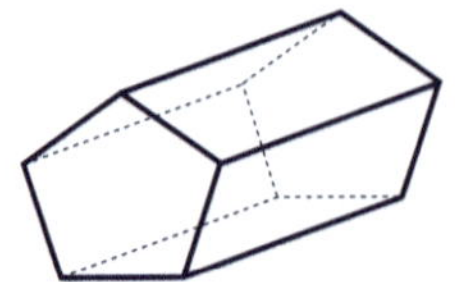

f

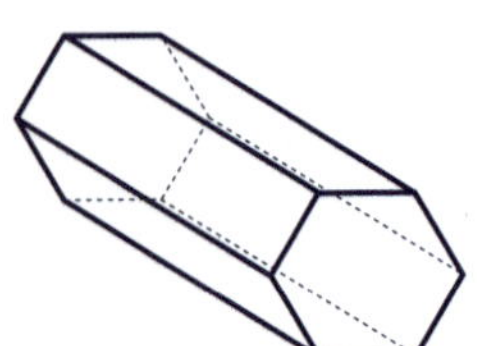

5 Draw the solid that can be made from each of the following nets.

a 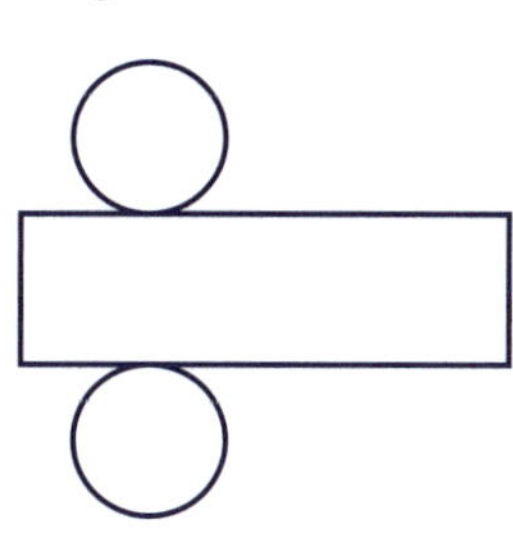

b 

c 

d 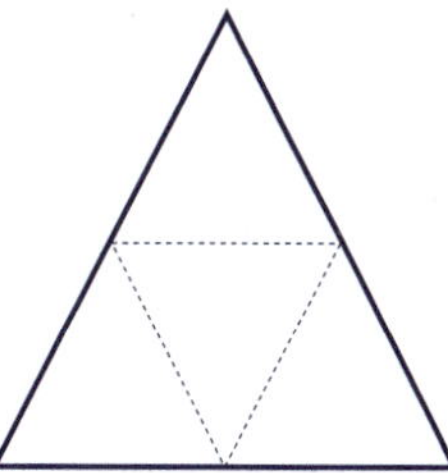

6 Draw an enlarged version of this net. Cut out the net and fold it to make an octahedron.

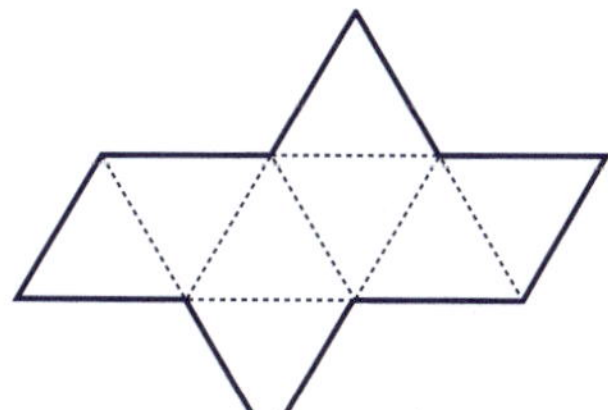

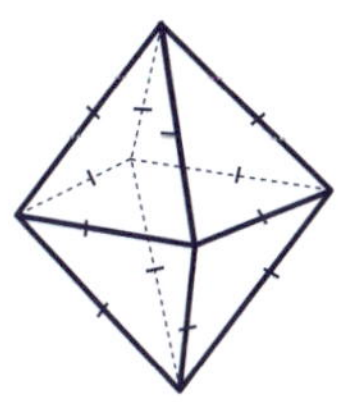

7 Which of the following nets can be folded to form a cube?

a 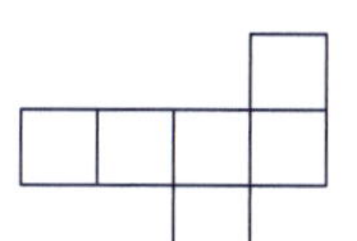

b 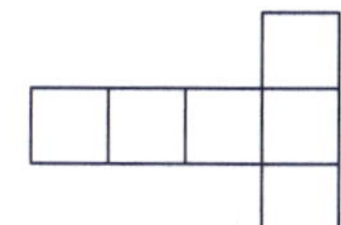

c 

d 

e 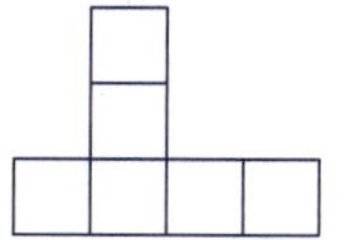

8 Which of the following nets can be folded to form a prism?

a 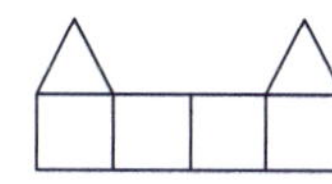

b 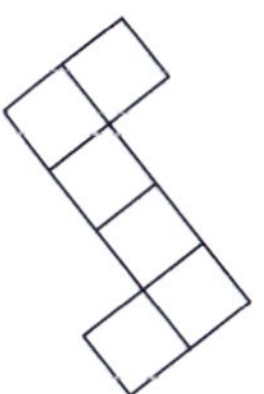

c 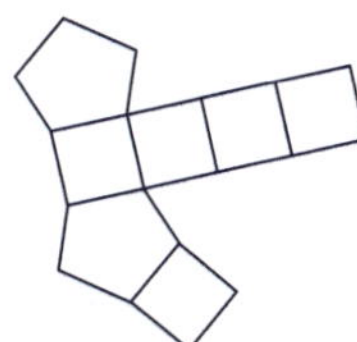

d 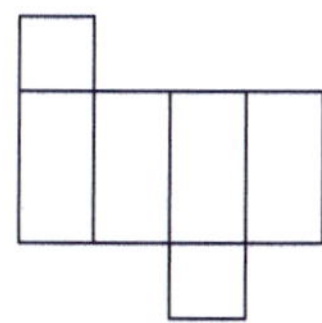

e 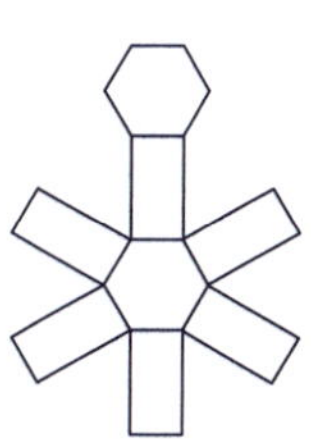

9 Which of the following nets can be folded to form a pyramid?

a 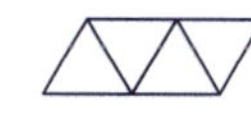

b 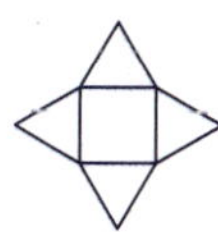

c 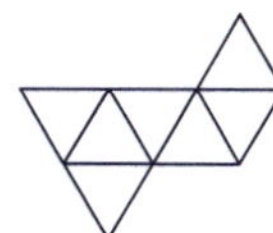

d 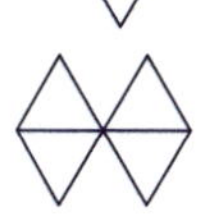

e 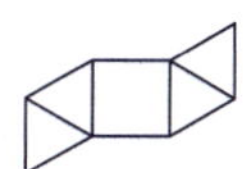

Challenge

This is a net for a 20-sided solid called an **icosahedron**. Investigate if it is possible to draw a different net that would fold to make the same solid. Describe what you found.

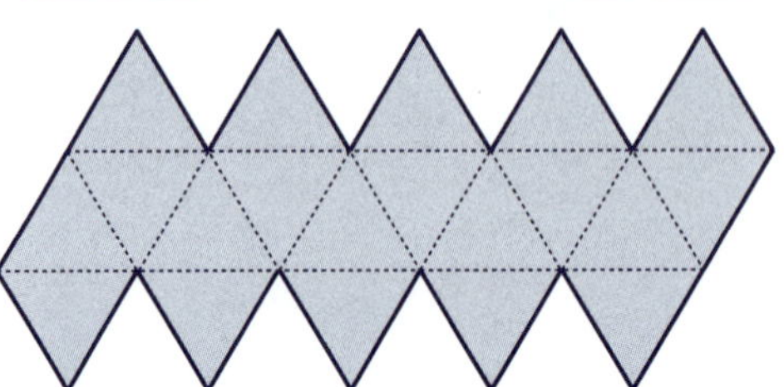

Lesson 4 The space inside

Earlier you learned that the **volume** of an object is the amount of space it occupies. The volume of a container can be found by packing the container with centicubes.

1 Count the number of centicubes to find the volume of each of these rectangular prisms.

a 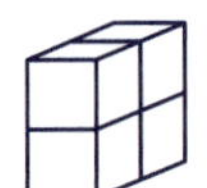

b 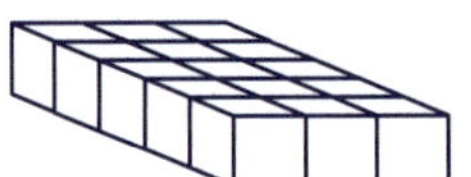

c 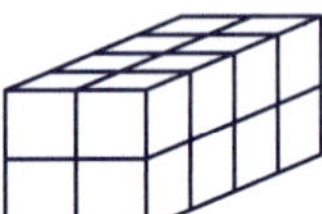

d 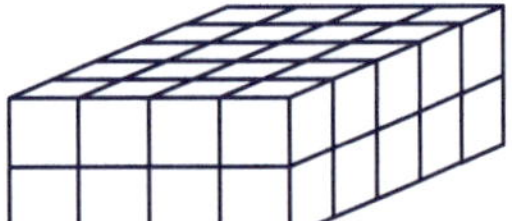

e 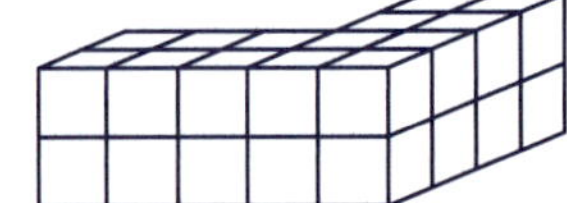

2 Count the number of centicubes needed to build each of these shapes and work out its volume. Remember to count the centicubes that cannot be seen.

a 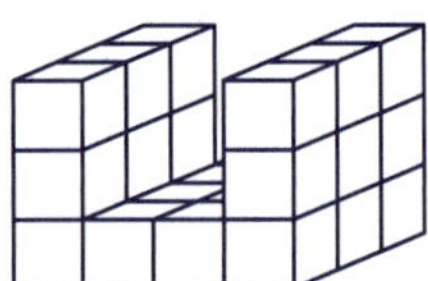

b 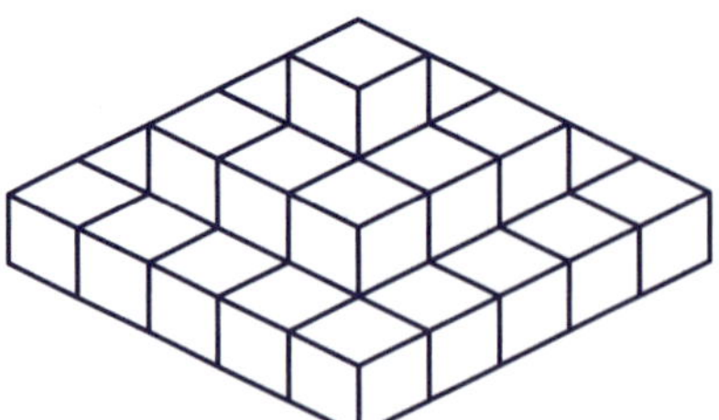

c 

d 

e 

3 Use the formula in the Help Box to find the volume of the following packages.

a

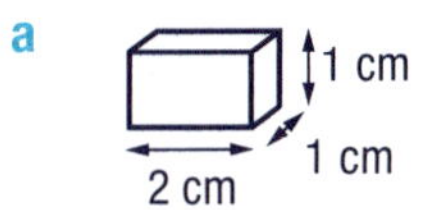

b

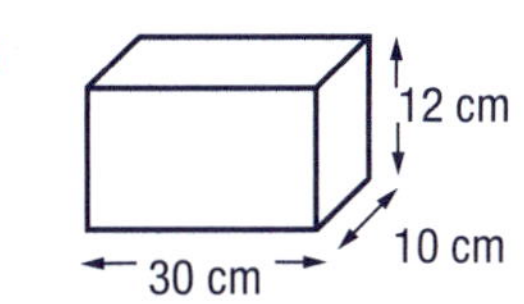

c

25 cm
10 cm
15 cm

d

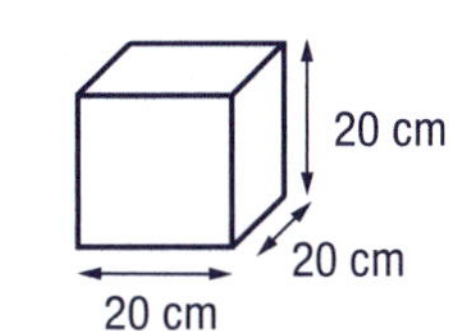

e

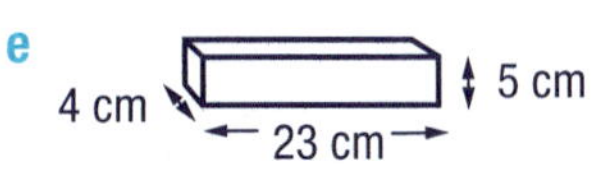

f

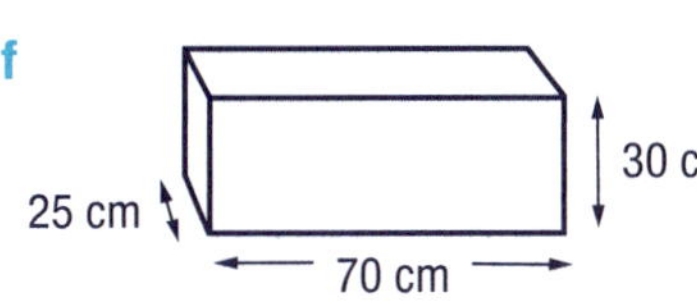

g

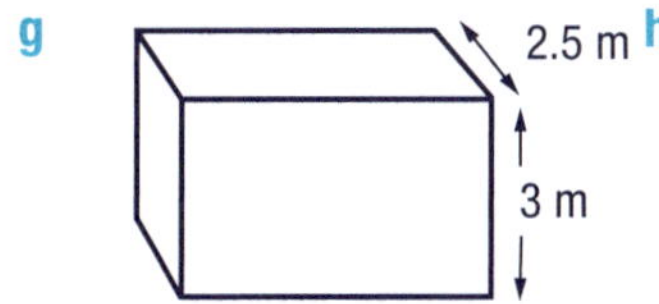

h

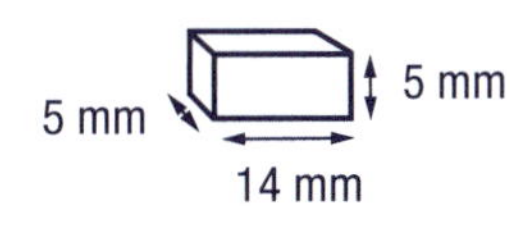

Help Box

It is also possible to calculate the volume of a rectangular prism by using the formula:

Volume (V) = length (l) × width (w) × height (h)

$$V = l \times w \times h$$

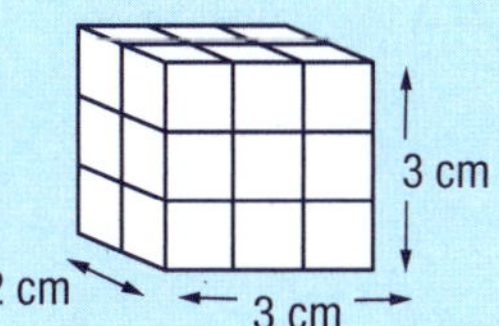

Therefore the volume of the rectangular prism
= 3 cm × 2 cm × 3 cm
= 18 cm^3

4 Calculate the volume of each of the following packages:

- **a** a shoe box with a length of 30 cm, a width of 17 cm and a height of 12 cm
- **b** a rectangular box of golf balls measuring 19 cm long, 5.5 cm wide and 5.5 cm high
- **c** a rectangular fishing basket with a length of 75 cm, a width of 28 cm and a height of 40 cm

Challenge

A sports shop needs to store 60 shoe boxes each with a length of 35 cm, width of 20 cm and height of 14 cm. What is the smallest amount of storage space needed to pack these shoe boxes?

Using the formula $l \times w \times h$ to find the volume of a prism is exactly the same as multiplying the area of the base by the height. The formula area of base (a) $\times$ height (h) can be used to find the volume of all **uniform** solids.

When **uniform** solids are cut perpendicular to the length they have uniform cross-sections. This means that all the cross-sections are the same size and shape. Prisms are uniform solids because they are made up of many layers of identical cross-sections.

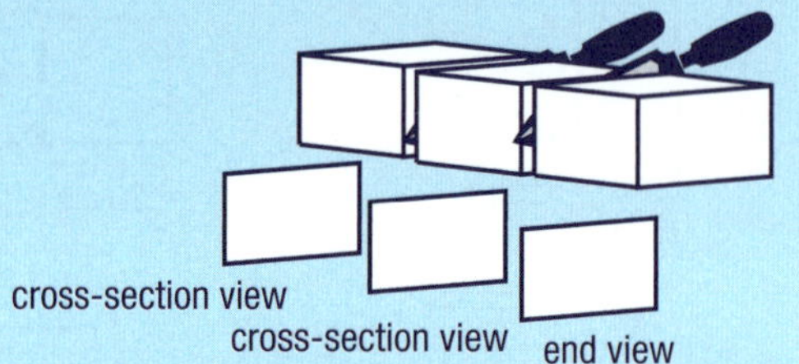

The formula for finding the volume of a prism or uniform solid is:

$$\begin{aligned}\text{Volume } (V) &= \text{length } (l) \times \text{width } (w) \times \text{height } (h)\\ &= \text{area of base } (A) \times \text{height } (h)\end{aligned}$$

5 Find the volume of these prisms.

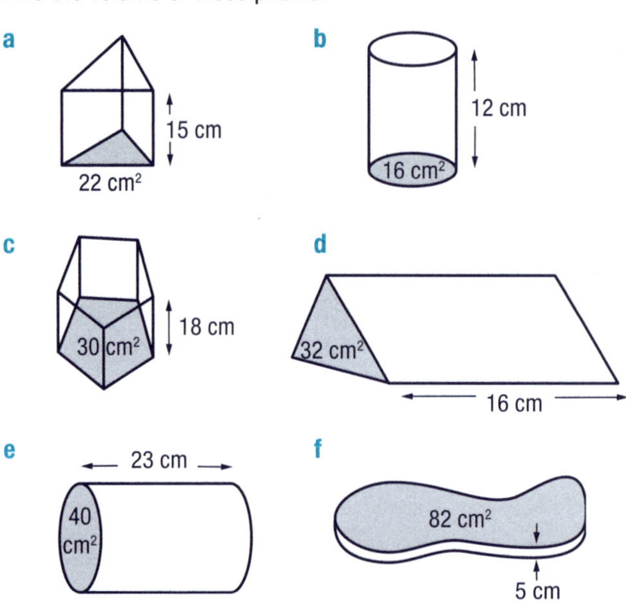

6 Use the formula to find the volume of:

- **a** a cylinder with a base area of 33 cm^2 and a height of 8 cm
- **b** a triangular prism 15 cm high with a base area of 24 cm^2
- **c** a 20 cm high uniform brick with an area at each end of 18 cm^2

Help Box

In Grade 6 you learned the formula for finding the area of a triangle:

$$\text{Area } (a) = \frac{\text{height } (h) \times \text{base } (b)}{2}$$

This diagram will remind you why the formula applies.

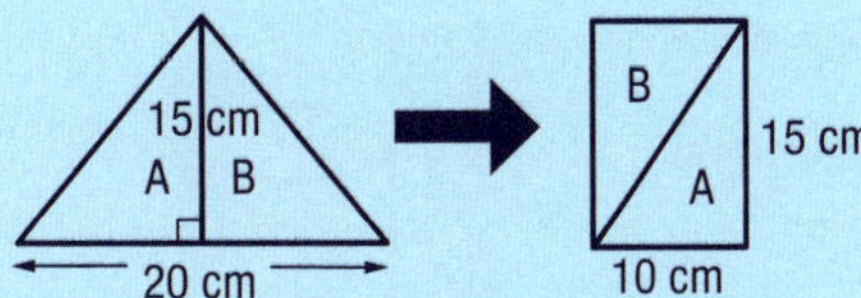

A triangular prism is a uniform solid. That means that its volume can be calculated using the formula $V = Ah$.

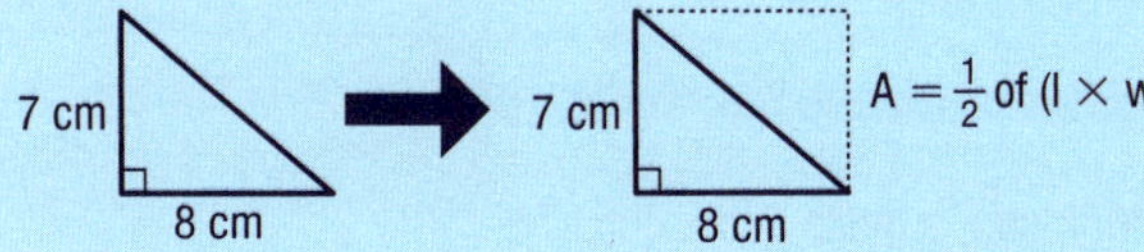

7 Use the formula to find the volume of the following triangular prisms.

a

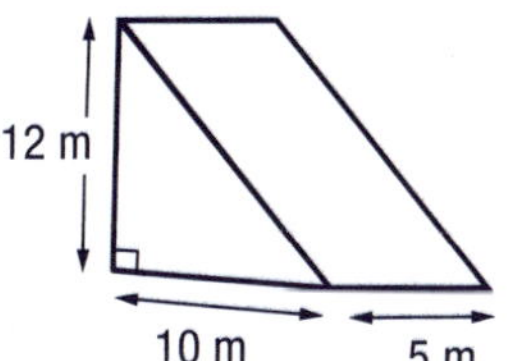

b

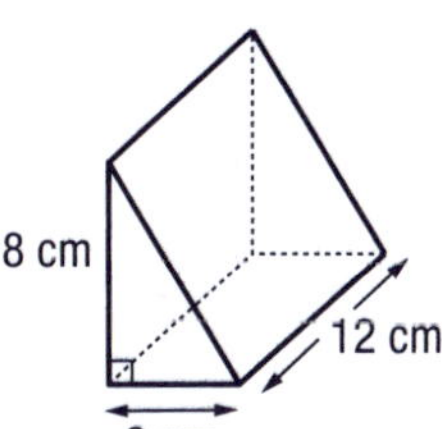

c

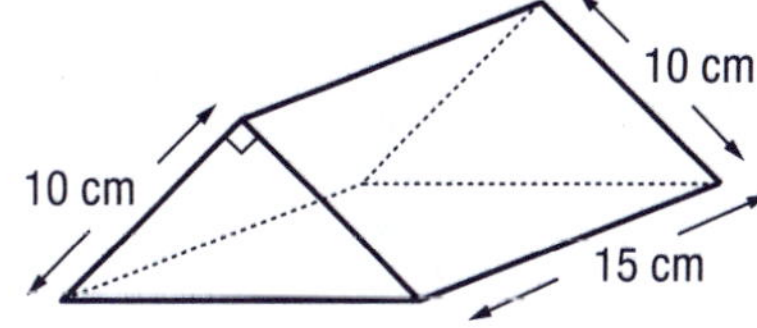

d

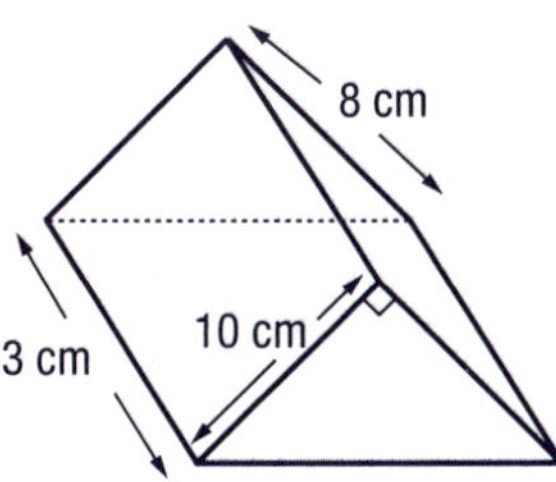

e

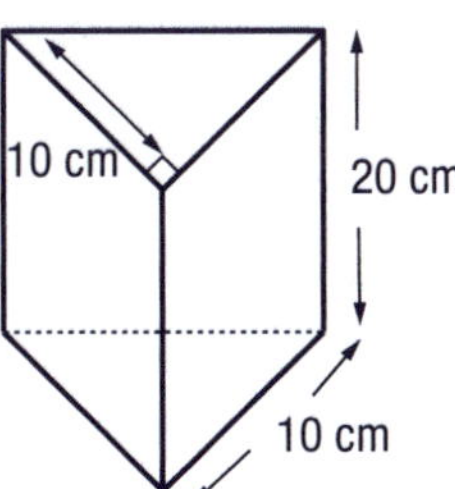

f

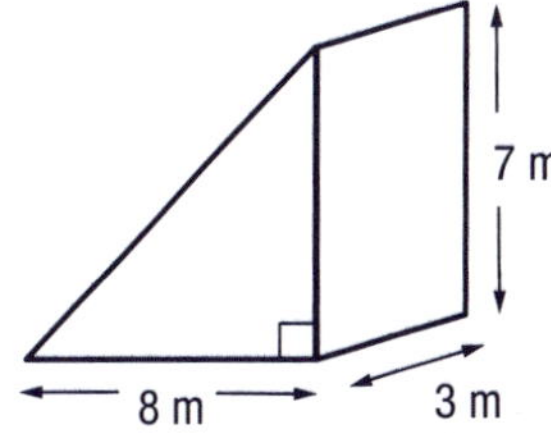

Help Box

Earlier you learned the formula for finding the area of a circle:

$$\text{Area } (a) = \pi r^2$$

A cylinder is a uniform solid, so its volume can be calculated using the formula $V = Ah$.

To calculate the volume of a cylinder when the base area is not given, we can use the formula:

$$\text{Volume } (V) = \text{Area } (\pi r^2) \times \text{height } (h).$$

For example:

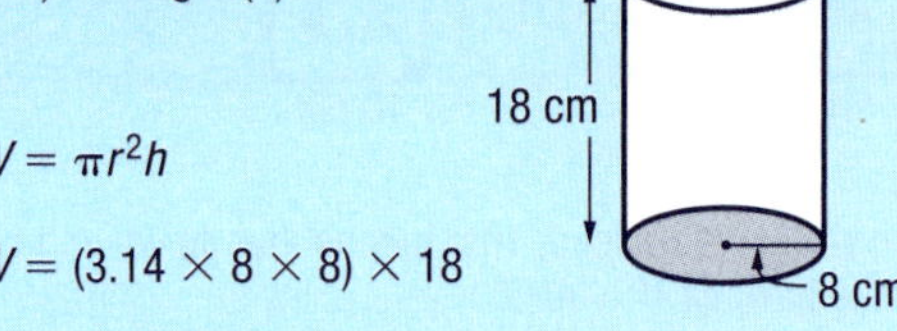

$$V = \pi r^2 h$$

$$V = (3.14 \times 8 \times 8) \times 18$$

$$= 3617.28 \text{ cm}^2$$

8 Find three different cylindrical containers such as a soft drink can, a glass and a cup.

- **a** Estimate how many centicubes it would take to fill each container.
- **b** Measure and write down in your workbook, the height and the diameter of each container
- **c** Use the formula to calculate the actual volume of the three containers.
- **d** What was the difference between each of your estimates and the actual volumes of the containers?

9 Use the formula to calculate the volume of the following cylinders.

a

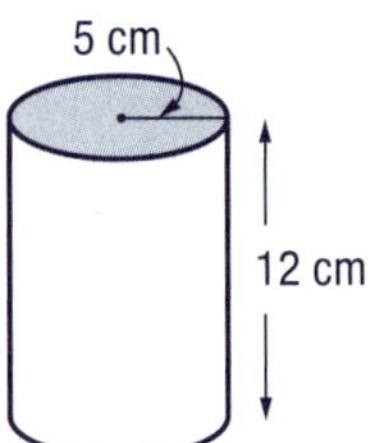

b

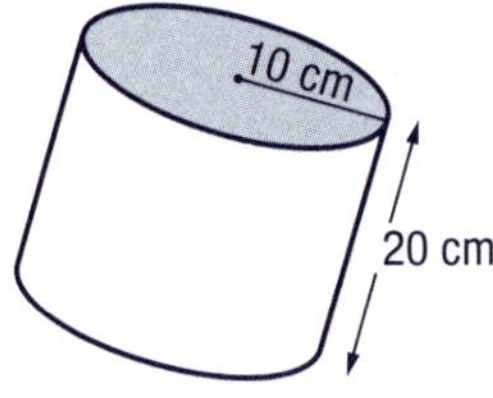

c

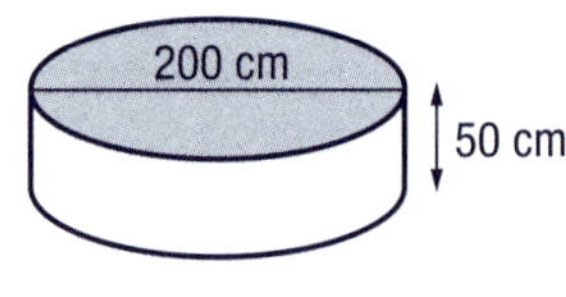

d

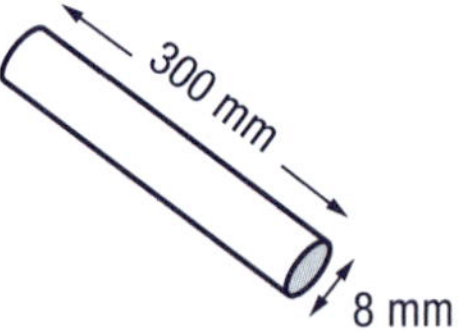

e

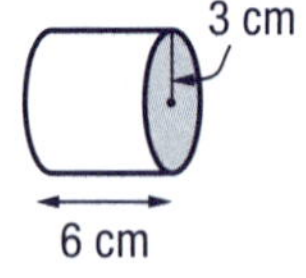

f

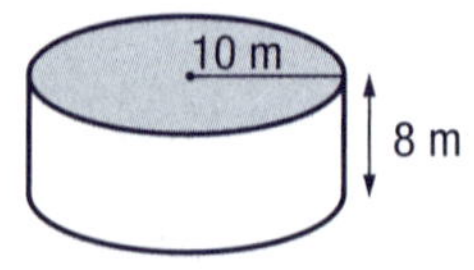

g

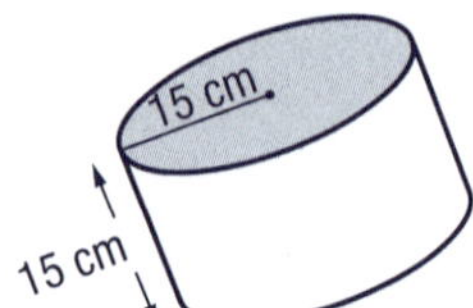

h

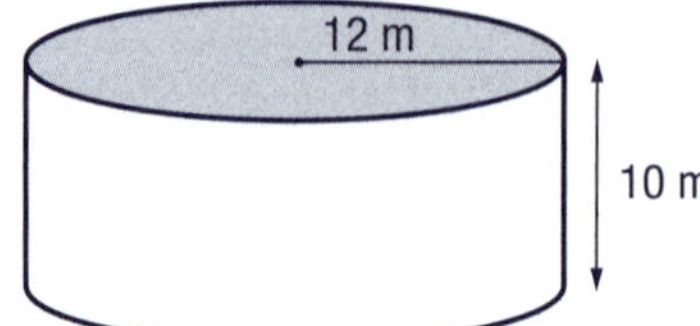

Lesson 5 Tapering solids

Not all solids are uniform solids. Some solids, such as pyramids and cones, taper to a point. When these solids are cut parallel to the base the shape of the cross-section is similar to but smaller than the base.

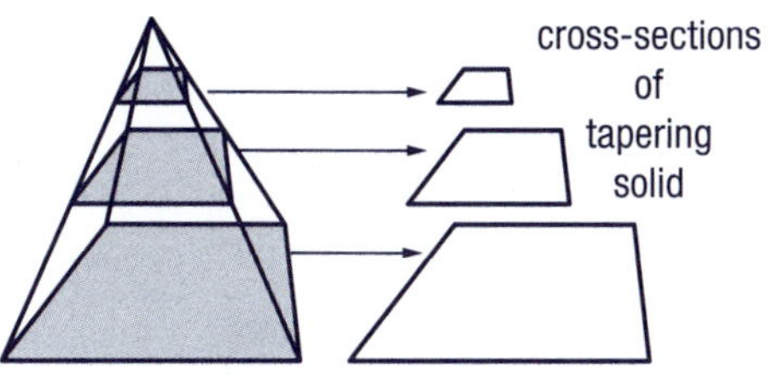

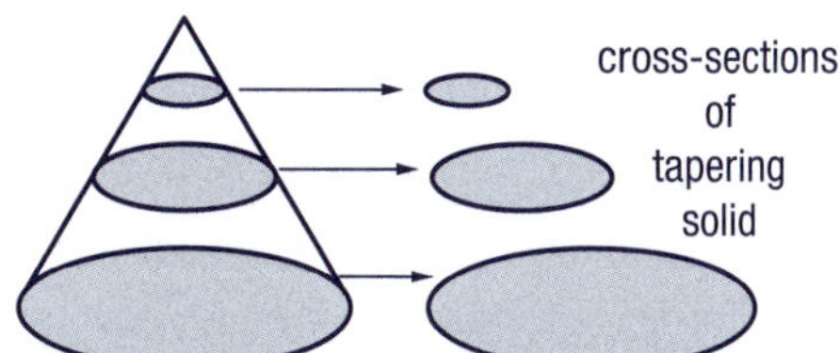

1 Work with a partner to complete the following investigation.

- Make a net for a square pyramid from stiff cardboard.
- Cut the net out. Fold and tape the edges to form the pyramid.
- Cut out and remove the square base.
- Estimate the number of centicubes that you would need to fill the pyramid.
- Line your pyramid with a plastic bag.
- Fill the lined pyramid with water.
- Empty the water into a measuring jug to find the capacity of the pyramid.
- Use the information to work out the volume of the pyramid.
- Copy the following table into your workbook and complete to record your findings.

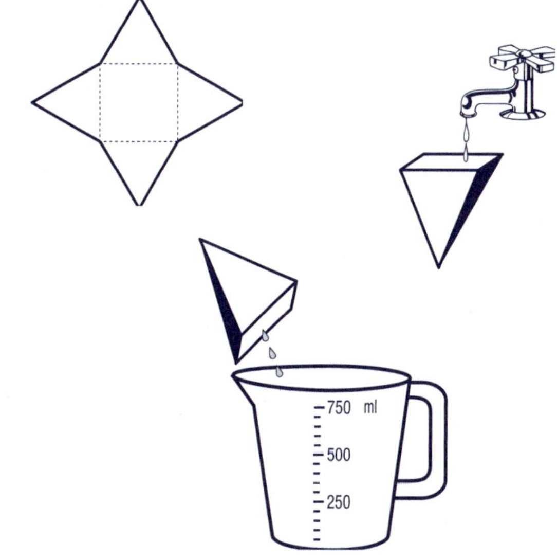

Dimensions of pyramid				Estimated volume (cm^3)	Capacity (mL)	Volume (cm^3)
Length of base	Width of base	Area of base	Height at apex			

2 Compare your results with some other students' and see if you can identify any relationship between the area of the pyramid's base, its height and its volume. Write a few sentences to describe what you found.

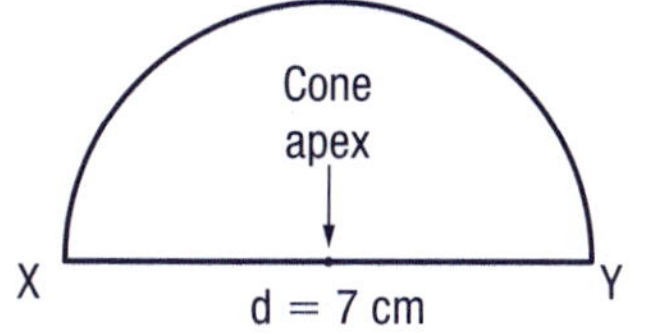

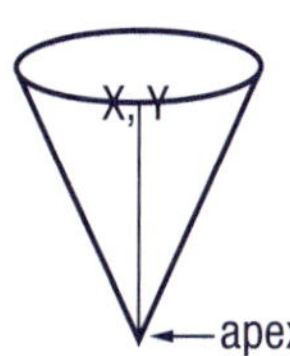

A **cone** is another example of a tapering solid. A cone can be made by drawing a sector of a circle as shown. The smaller the sector the sharper the cone will be at its apex.

3 Work with a partner to investigate the volume of cones.

- Draw a sector of a circle onto paper.
- Cut the sector out and tape it to form a cone.
- Estimate the number of centicubes that you would need to fill the cone.
- Line your cone with a plastic bag.
- Fill the lined cone with water.
- Empty the water into a measuring jug to find the capacity of the cone.
- Use the information to work out the volume of the cone.
- Copy the following table into your workbook and complete to record your findings.

Dimensions of cone		Estimated volume (cm^3)	Capacity (mL)	Volume (cm^3)
Area of base ($\pi d \div 2$)	Height at apex			

4 Compare your results with some other students' and see if you can identify any relationship between the area of the cone's base, its height and its volume. Write a few sentences to describe what you found.

The formula for calculating the volume of *all* tapered solids is:

$$\text{Volume } (V) = \tfrac{1}{3}(\text{Area of base}) \times \text{height } (h)$$

For example: $V = \frac{1}{3}$ of $A \times h$

$V = \frac{1}{3} \times 36 \times 10$

$= 120 \text{ m}^3$

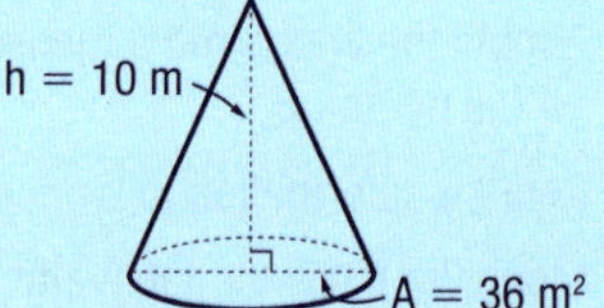

5 Use the formula to calculate the volume of the following tapered solids.

a

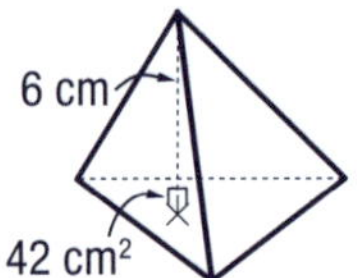

b

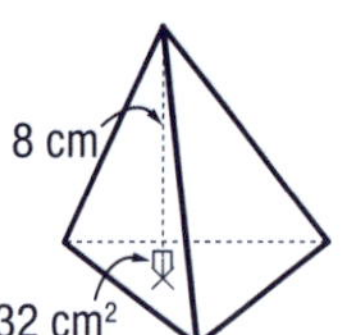

c

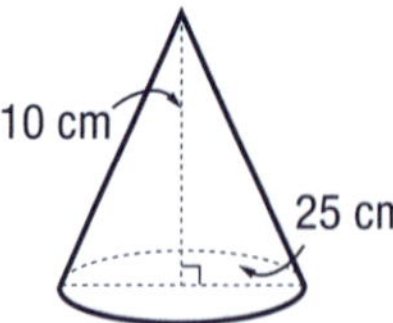

d

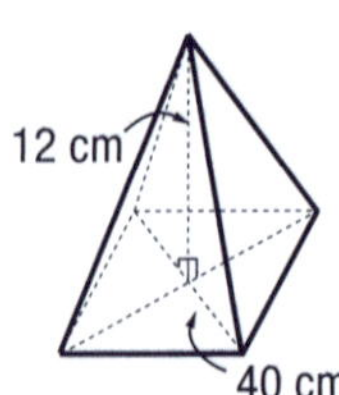

e

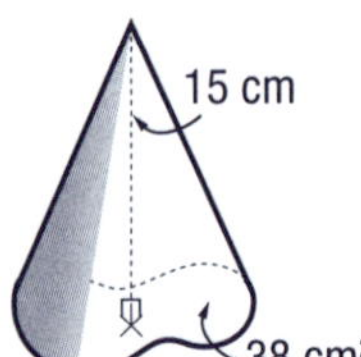

f

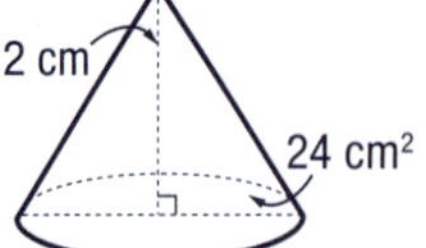

g

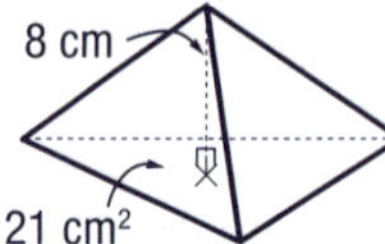

h

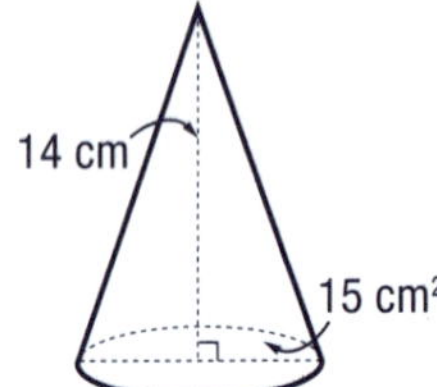

Challenge

The volume of a cone is approximately 125 cm^3. The height of the cone is 14 cm. What is the radius of the circle that forms the base of the cone?

Lesson 6 Problem solving with volume

A company making sports drinks has the option of packaging their product in small plastic-coated cardboard prisms or in aluminium cans.

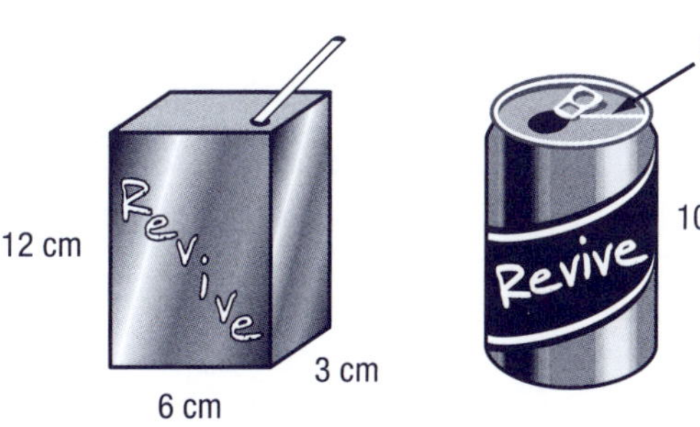

1
- **a** Calculate the volume of each container.
- **b** How many millilitres would each container hold?
- **c** If the company intends charging K2 per container, which container would be most profitable (excluding the cost of packaging)?
- **d** Draw and label two sketches to show how many of each container would fit onto a supermarket shelf measuring 30 cm by 18 cm (draw the containers only one stack high).
- **e** Write a few sentences describing which container the company should select and why.

2 Calculate the volume of the following containers using the appropriate formula in each case.

a

7 cm
15 cm
40 cm

b

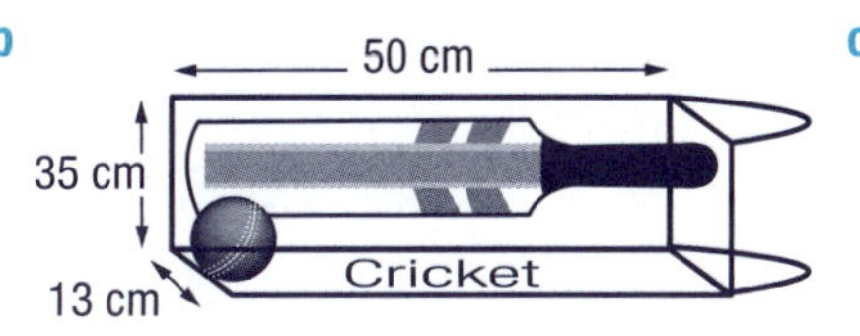

c

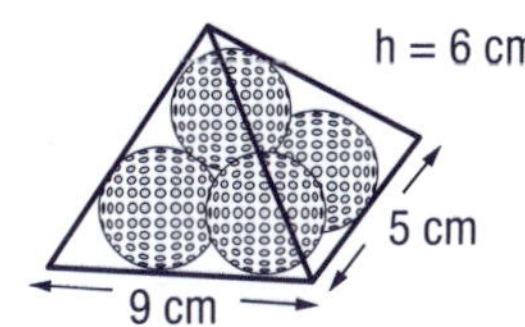

d

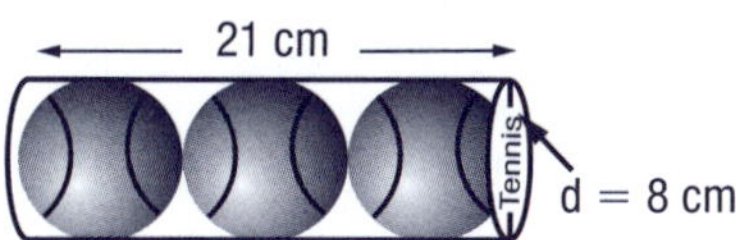

e

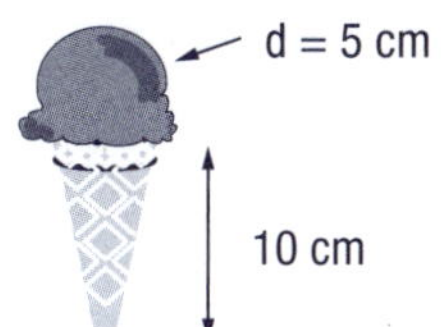

f

3 Which container of energy drink is better value?

- **A** a rectangular prism measuring 12 cm by 8 cm by 4 cm costing K1.80 or
- **B** a can with a diameter of 8 cm and a height of 12 cm costing K2.50

4 Patricia wants to buy a container of marbles for a gift. She has the choice of buying a cardboard cone or a square pyramid. The cone has a radius of 4 cm and is 12 cm tall. The pyramid has a side length of 5 cm and is 10 cm tall.

- **a** Draw each container and label the dimensions.
- **b** Calculate the volume of each container to determine which is larger.

5 Draw and label the dimensions of three different sized rectangular prisms that each has a volume of between 40 cm^3 and 50 cm^3.

6 Draw and label the dimensions of two different sized cylinders that each has a capacity of between 300 mL and 400 mL.

7 Draw and label the dimensions of two different tapered solids that each has a volume between 80 cm^3 and 100 cm^3.

8 Calculate the value of x in each of the following to the nearest whole number.

a

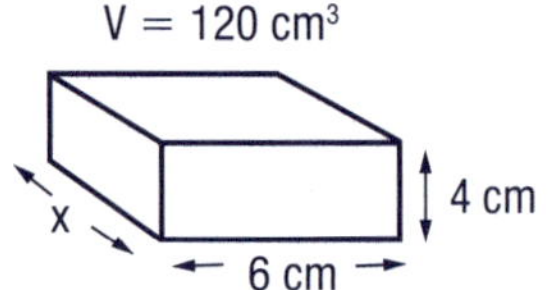

b

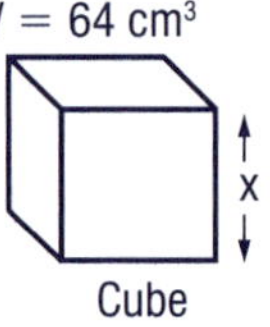

c

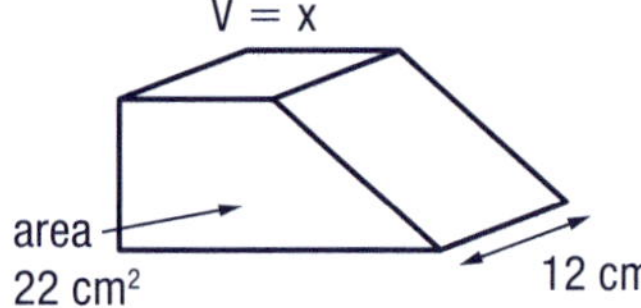

d

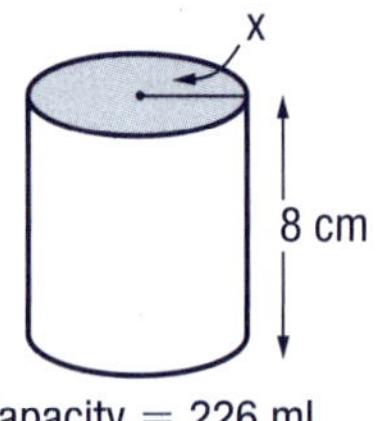

e

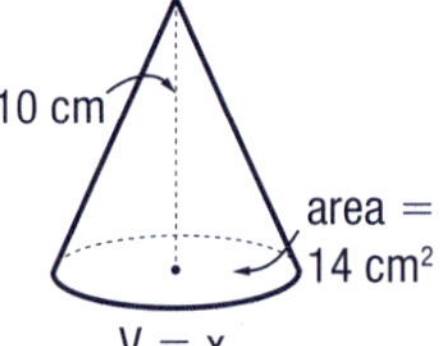

f

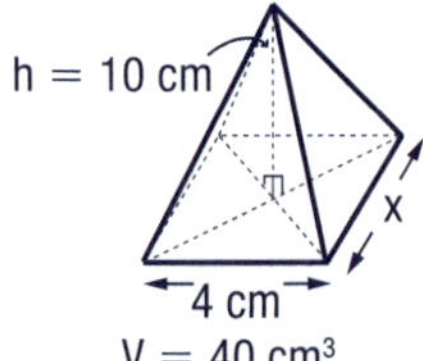

Challenge

A cone with a diameter of 6 cm and a height of 7 cm is full of liquid. If all the liquid is poured from the cone into a can of the same diameter, what will be the height of the liquid in the can?

Lesson 7 Packing efficiently

1 Sam has to stack 36 cylinders containing sports balls onto a shelf. The cylinders have a diameter of 8 cm and are 23 cm high. There is a space of 50 cm between the shelf and the one above.

- **a** What is the volume of each cylinder?
- **b** In your workbook draw at least three different ways that Sam can stack the cans.
- **c** Which model of stacking will require the least shelf space?
- **d** Calculate the total shelf space taken up by the 36 cylinders using that model of stacking.
- **e** While this model is the most efficient for saving shelf-space, how practical is it for taking the cylinders off the shelf? Explain your answer.

2 Sam also needs to stack 20 boxes containing swimming goggles onto a shelf with the same space available as in Question 1.

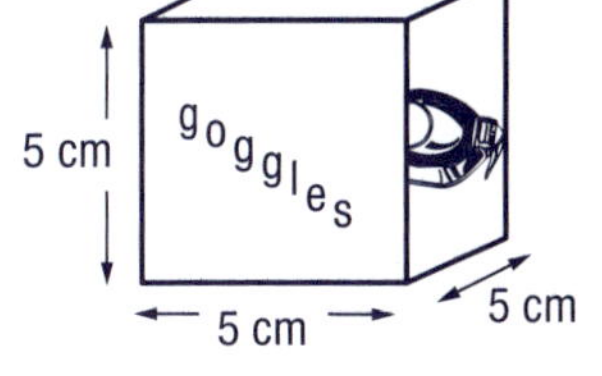

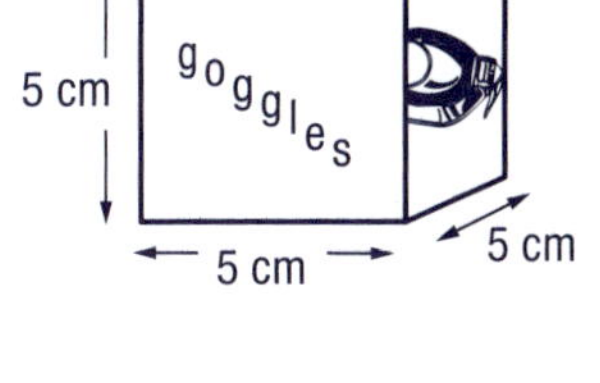

- **a** What is the volume of each box?
- **b** Draw three different ways that Sam could stack the 20 boxes.
- **c** Which model of stacking will require the least shelf space?
- **d** Calculate the total shelf space taken up by the 20 boxes using that model of stacking.
- **e** While this model is most efficient for saving shelf-space, how practical is it for taking the boxes off the shelf? Explain your answer.

3 Sam also has to stack 16 family tennis sets onto a similar shelf. The set of four racquets are in a box with a square base.

- **a** What is the geometrical name for this shape of box?
- **b** What is the volume of each box?
- **c** How many different ways could Sam stack the 16 boxes in the space available?
- **d** Which model of stacking will require the least shelf space?
- **e** Calculate the total shelf space taken up by the 16 boxes using that model of stacking.
- **f** While this model is most efficient for saving shelf-space, how practical is it for taking the boxes off the shelf? Explain your answer.

4 A stallholder wants to design three different containers to hold the following items:

- 24 cubes each with a side length of 3 cm
- 16 rectangular prisms each 10 cm long, 5 cm wide and 6 cm high
- 24 cylinders each with a diameter of 8 cm and a height of 12 cm

What might be the dimensions of each of the three containers?

5 Energy drinks are sold in boxes measuring 11 cm high, 6 cm wide and 5 cm deep. Draw and label the dimensions of a carton that could be used to pack six boxes of energy drink so that there is minimal wastage of space.

6 Another brand of energy drink is sold in cans with a diameter of 7 cm and a height of 8 cm. Draw and label the dimensions of a carton that could be used to pack six cans of this energy drink so that there is minimal wastage of space.

7 The local sports club wants to install some seating along a 10 m boundary of the athletic track. The committee has decided each seat needs to be 40 cm × 40 cm (1600 cm^2).

- **a** What is the maximum number of seats that can be installed if only a single row of seats is to be built?
- **b** If a second row of seats was installed behind the first row including an additional 40 cm space for leg room between the rows, what would be the total area taken up by the seated spectators?
- **c** Draw a plan and label the dimensions to show how you calculated your answers.

Challenge

Which option would require the lesser area and accommodate the most spectators:

- a single row of spectators seated along a 25 metre length of the boundary or
- three rows of spectators seated along the 10 metre boundary with 40 cm leg space allowed for each row?

8 At the swimming carnival spectators were seated 24 in every row, and there were 8 rows of spectators.

a How many spectators were at the carnival?

b If each seat takes up 40 cm $\times$ 40 cm and there is leg room of 40 cm provided for each row, what was the total area occupied by the spectators?

c How else might the seats have been arranged to accommodate that many people?

d What area would this seating arrangement use?

9 List some factors that need to be considered when planning seating arrangements for spectator sports.

Lesson 8 Bits and pieces

1 A company that makes fishing baskets for adults wants to produce a child's version. The length and height of the child's basket will be $\frac{2}{3}$ of the adults' basket. The width will be $\frac{3}{4}$ of the adults' size. The adult bag is shown here. Draw the child's basket and label its dimensions.

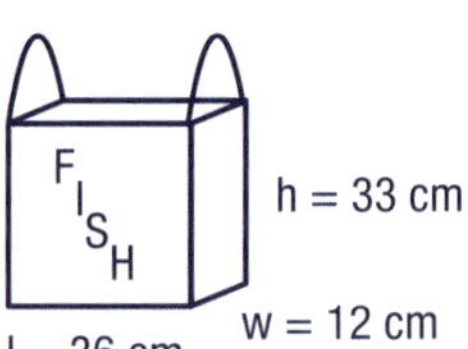

2 Reduce the size of each of the following measurements by $\frac{1}{4}$.

a 20 cm b 28 cm c 12 m d $16\frac{1}{2}$ mm e $17\frac{1}{2}$ m

f $9\frac{2}{3}$ km g $15\frac{3}{4}$ cm h $11\frac{3}{8}$ m

3 A company makes 'first prize' ribbons for competitions.

a How many ribbons can they make from $12\frac{1}{4}$ metres of red ribbon if each small ribbon is $\frac{1}{5}$ of a metre in length?

b How much ribbon is wasted?

Challenge

What fraction of a metre should the company cut the ribbon in order to make the same number of ribbons from the $12\frac{1}{4}$ metres but not have any wastage?

4 Rodney trains for $13\frac{3}{4}$ hours each week during the holidays. When school resumes he has been told that the amount of time available to train will be halved. How long will training be then?

5 A stadium holds 1200 spectators when full. When the game begins the stadium is $\frac{3}{8}$ full. For every hour of play, an average of 80 additional spectators arrive. Every hour $\frac{1}{50}$ of the original group of spectators leave. How many spectators will be in the stadium after four hours?

6 Copy the following numbers into your workbook and insert the symbol $>$ or $<$ to make a true statement.

a $\frac{1}{4} * \frac{5}{12}$ b $\frac{3}{9} * \frac{3}{8}$ c $\frac{3}{10} * \frac{3}{4}$ d $\frac{2}{3} * \frac{4}{7}$

e $\frac{1}{2} * \frac{10}{17}$ f $\frac{4}{5} * \frac{17}{20}$ g $\frac{5}{6} * \frac{7}{8}$ h $\frac{2}{7} * \frac{1}{3}$

7 A stallholder at the market sells mugs of cordial. Each mug holds $\frac{1}{3}$ litre. He starts the day with 3½ litres of cordial.

a How many mugs can be filled from the container?

b When the stallholder sells 7 mugs of cordial, how many litres will be left in the container?

c How many mugs must he sell before the level in the container falls below ¾ of a litre?

d If the stallholder decides to put $\frac{3}{8}$ of a litre in each mug instead of $\frac{1}{3}$ of a litre, how would the answers to parts a, b and c change?

8 Calculate:

a $\frac{2}{3} + \frac{4}{7}$ b $\frac{2}{5} + \frac{7}{8}$ c $\frac{7}{10} - \frac{1}{2}$ d $\frac{5}{9} - \frac{1}{3}$

e $\frac{3}{8} \times \frac{9}{10}$ f $\frac{13}{15} \times \frac{1}{10}$ g $\frac{3}{4} \div \frac{3}{6}$ h $\frac{8}{11} \div \frac{1}{4}$

9 The local swimming hole was $\frac{3}{8}$ full before heavy rain fell for three days and raised the water level by $\frac{1}{16}$ each day.

a What fraction of the swimming hole was full on each of those three days?

b What fraction of the swimming hole was empty after the three days?

c If the swimming hole had a total capacity of 120 000 L, how many more litres are needed to fill it?

Challenge

A water tank holds 400 L of water but it has a small crack in the bottom. Every night the tank's level drops by $\frac{1}{25}$ L. If no-one uses any water from the tank, how many nights will it take for the tank to empty?

10 A swimming coach tells his student to swim 12 laps of the pool as a warm-up routine. Each lap is 50 metres long.

a How many laps is equal to $\frac{1}{3}$ of the warm-up routine?

b What fraction of the warm-up routine is completed at 9 laps?

c What distance will the student have swum when she is $\frac{2}{3}$ through the warm-up?

11 Spectators at a soccer match are seated in bays. Two bays are shown here.

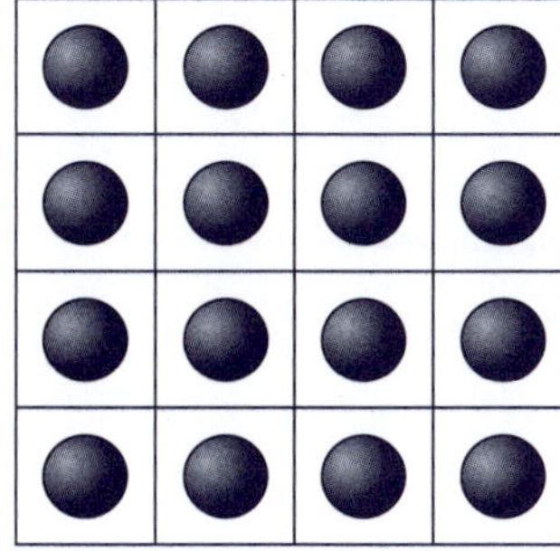

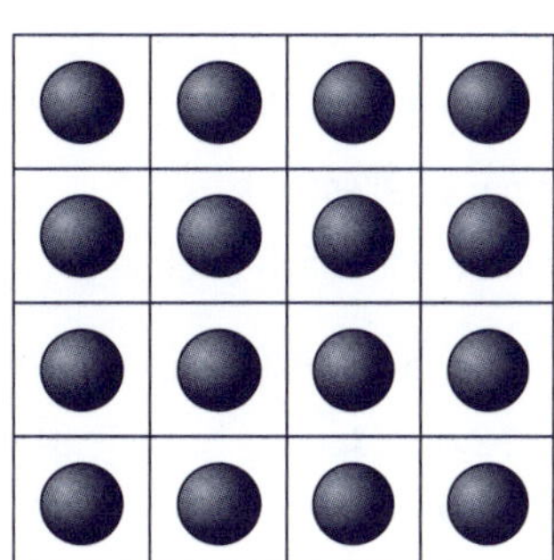

a How many spectators can be seated in one bay?

b How many spectators can be seated in $2\frac{1}{2}$ bays?

c If there is a total of 8 bays at the soccer and each bay is only $\frac{3}{8}$ full, how many spectators are at the match?

d How many additional spectators would be needed if the stadium manager wants to have the stadium $\frac{3}{4}$ full in order to make a profit?

12 In a swimming squad of 40 students, there are 24 boys. One-third of the boys and half the girls are in the freestyle event. How many students in the squad are in the freestyle event?

13 There are 47 students in the athletics team of which 27 are girls. Two-thirds of the girls and one-quarter of the boys are in the 200-metre sprint. How many students from the team are in the 200-metre sprint?

Lesson 9 Fractions and decimals

In Grade 6 you learned that fractions and decimals are different ways of describing the same portion of a unit.

1 Copy the table into your workbook and complete it to refresh your memory.

Division algorithm	Fraction	Convert to decimal denominator	Decimal
$3 \div 4$	$\frac{3}{4}$	$\frac{75}{100}$	0.75
$1 \div 2$			
$7 \div 10$			
$4 \div 5$			
$21 \div 100$			
$13 \div 20$			
$3 \div 8$			
$1 \div 3$	$\frac{1}{3}$	$\frac{33}{100}$	0.33

2 List five fractions with a numerator other than 1 with each having a different denominator.

a Order your fractions from smallest to largest.

b Write the decimal equivalent for each fraction to one decimal place.

3 Solve the following problems by first converting all the numbers to common fractions.

a $12\frac{1}{2} + 0.25$ b $2.8 + 3$ c $8\frac{1}{4} - 5.2$ d $4.3 - 2\frac{1}{2}$ e $5.7 \times 6\frac{3}{10}$

f $2\frac{4}{10} \times 4.06$ g $3.25 \div \frac{1}{2}$ h $3\frac{3}{8} \div \frac{1}{4}$ i $6.25 \div 1\frac{1}{4}$ j $6\frac{7}{20} \times 4.2$

4 Redo the problems in question 3 by first converting all the numbers into decimal numbers.

5 Calculate the cost of the following.

a $4\frac{1}{3}$ litres of energy drink at K1.80 per litre

b $12\frac{1}{4}$ litres of petrol at K3.20 per litre

c $3\frac{1}{2}$ kg of fish at K2.50 per kilogram

d $6\frac{3}{4}$ m of ribbon at K0.75 per metre

6 A professional athlete donated $\frac{11}{20}$ of his daily winnings of K80.40 to a charity. How much did the athlete donate for the day?

7 About $\frac{2}{3}$ of the human body is comprised of water. How much water might be contained in a person's body if they have the following mass?

a 39.3 kg b 45.9 kg c 66.3 kg d 81.6 kg

8 How many ribbons 0.25 m in length can be cut from a piece measuring:

a 5 m long b $2\frac{1}{4}$ m long c $4\frac{1}{4}$ m long d $10\frac{3}{4}$ m long e $9\frac{3}{4}$ m long f $7\frac{1}{2}$ long?

Challenge

Write two word problems of your own involving fractions and decimals in the same problem. One problem should require multiplication, and the other should require division to calculate a solution. Remember to work out the answers as well. Now challenge a friend to solve your problems.

9 What fraction is:

a 4 hours of 32 hours b 1.5 days of the whole weekend c 125 mL of 2.5 L

d 2.5 m of 15 m e 4.2 m of $16\frac{4}{5}$ m f 7.5 min of $4\frac{1}{2}$ hours?

10 What is the total volume of a box if:

a 5.2 cm^3 is $\frac{1}{3}$ of the total b 5.2 cm^3 is $\frac{1}{4}$ of the total

c 6.4 cm^3 is $\frac{7}{8}$ of the total d 6.4 cm^3 is $\frac{1}{8}$ of the total

e 28.4 mm^3 is $\frac{1}{4}$ of the total f 28.4 mm^3 is $\frac{3}{4}$ of the total

g 3.84 cm^3 is $\frac{2}{3}$ of the total h 3.84 cm^3 is $\frac{1}{3}$ of the total?

Learning Unit 4 Additional Learning, Revision and Assessment

Strand: Number and Application

Indices	Outcome 8.1.8	Use integer indices and fractional indices where the answers are rational

Lesson 1: Additional learning
Focus: Recognising square numbers and finding prime factors

Lesson 2: Revision
Volume of uniform and non-uniform solids
Nets
Packaging
Fractions and decimals in problem solving
Indices

Lesson 3: Assessment Task 1
Practical investigation—group work
Assessment Task 1 will assess learning outcomes 8.2.7, 8.2.8, 8.2.13 and 8.2.6.

40 marks

Lesson 4: Assessment Task 2
Test of basic skills and routine applications
Assessment Task 2 will cover the learning outcomes from across the topic 'And the Winner is . . .'
The test will assess the extent to which students can:

- Demonstrate an understanding of the mathematical concepts
- Correctly choose and apply mathematical techniques to solve problems

60 marks
Total 100 marks

Lesson 1 Additional learning

Square numbers and prime factors

In Grade 7 you learned how mathematicians use index notation as a shortcut way of writing some multiplication facts. This lesson will build on that previous learning.

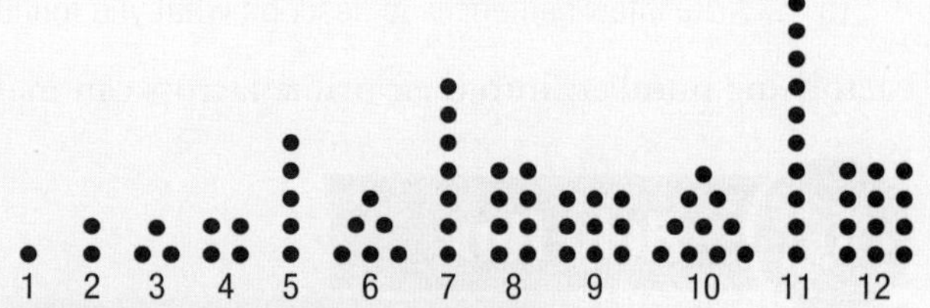

Some numbers can be arranged into a square pattern, while some can be arranged to form a triangle. Other numbers can only be formed into lines or rectangles. The diagram on the right shows how different numbers are related to each other because they can be represented in similar ways.

Help Box

4 'to the power of' 2 is written as 4^2.

It is a shortcut way of writing 4×4.

Sometimes this is called '4 squared'.

Here are the numbers from the diagram above that can be arranged into a square. Note the relationship between square numbers and how we calculate the area of a square.

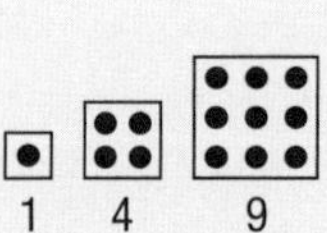

1 Copy and complete the following table about the first 10 square numbers.

Sequence in pattern	1st	2nd	3rd	4th	5th	6th	7th	8th	9th	10th
Calculation	1 × 1	2 × 2	3 × 3							
Product	1	4	9							

2 Copy into your workbook and complete:

a the 15th square number is .... × .... = ...

b the 20th square number is .... × .... = ...

c the 40th square number is .... × .... = ...

d the 100th square number is .... × .... = ...

e the 200th square number is .... × .... = ...

3 Calculate:

a 25^2 b $\sqrt{100}$ c $5^2 + 4^2$

d $\sqrt{49}$ e $\sqrt{81} + 4^2$ f $9^2 + \sqrt{4}$

g $\sqrt{64} + 17$ h $25^2 - \sqrt{9}$ i $\sqrt{36} - 5^2$

j $3^2 + \sqrt{121}$

Help Box

The opposite of squaring a number is finding the **square root** of a number. This means finding the number that was squared to produce the larger number. The symbol for square root is $\sqrt{\ }$.

For example, the square root of 64 can be written as $\sqrt{64}$.

The number that has been multiplied by itself to produce 64 is 8.

Therefore, the square root of 64 is 8 ($\sqrt{64} = 8$).

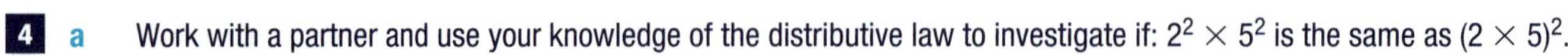

4 a Work with a partner and use your knowledge of the distributive law to investigate if: $2^2 \times 5^2$ is the same as $(2 \times 5)^2$.

b Check your idea to see if:

$8^2 \times 3^2$ is the same as $(8 \times 3)^2$

$4^2 \times 7^2$ is the same as $(4 \times 7)^2$

$2^2 \times 9^2$ is the same as $(2 \times 9)^2$

c Write a few sentences to describe what you found.

Factorising numbers into their prime factors can make it much easier to work with index notation.

Help Box

A factor tree finds two factors at a time until all the factors are prime numbers. For example, we can use the factor tree method to:

a find the prime factors of 156

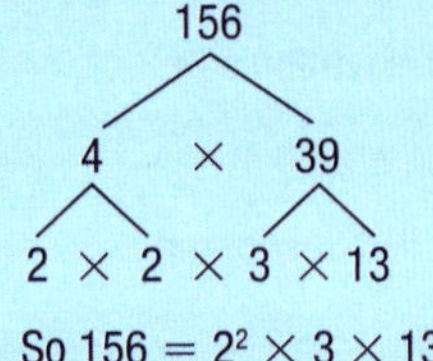

So $156 = 2^2 \times 3 \times 13$

b find the prime factors of 144

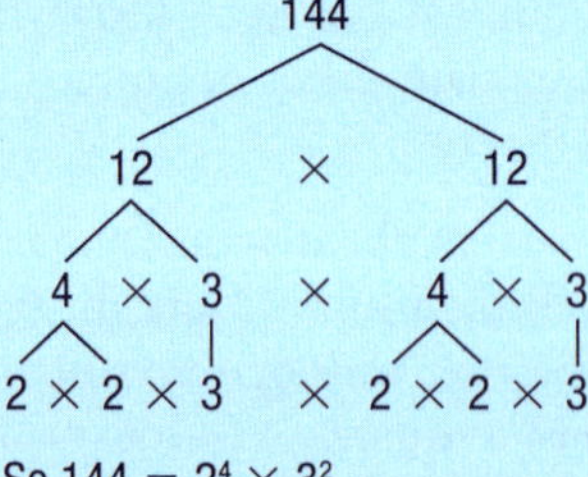

So $144 = 2^4 \times 3^2$

5 Copy and complete the following factor trees to find the prime factors.

a

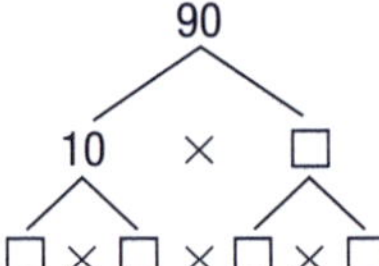

b

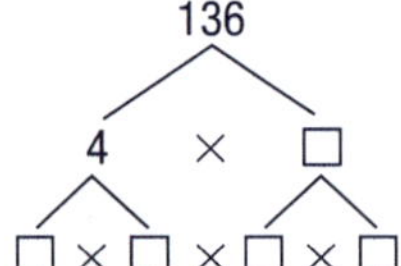

c

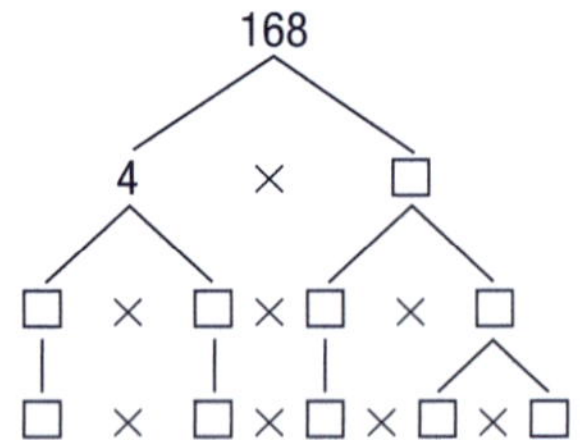

6 Draw factor trees to find the prime factors of the following numbers:

a 18 b 50 c 60 d 126

e 254 f 684 g 195 h 156

7 Write the following numbers as a product of prime factors. The first one has been done as an example.

a $42 = 7 \times 3 \times 2$ b $30 =$ c $24 =$

d $45 =$ e $63 =$ f $125 =$

8 Copy the table below into your book and complete it.

Number	Prime factors	Index notation
	$3 \times 3 \times 5 \times 5$	$3^2 \times 5^2$
	$2 \times 2 \times 6 \times 6$	
216		
		$2 \times 10^2 \times 3^2$
2600		

Challenge

Write the number 4 154 245 000 as a product of prime numbers.

Lesson 2 Revision

Check your understanding of this topic using the following exercises.

Unit one

1 Simplify the following ratios:

a 6:20 b 12:60 c 55:10

d 45:18 e 15:105 f 21:49

2 Use the formula for calculating experimental probability to estimate how many times a card that is thrown into the air 80 times will land 'playing side up'.

3 Calculate the probability that a normal dice, when rolled, will land on 6.

4 This table shows the results when some students spun a spinner with four sectors. Calculate the experimental probability of each sector being spun.

Pattern	A	B	C	D
Number of occurrences	25	15	55	5

5 Calculate the theoretical probability of:

a drawing a queen from a shuffled pack of cards

b drawing a black card from a shuffled pack of cards

c drawing a number card from a shuffled pack of cards

d rolling an odd number with a six-sided dice

e rolling number 5 with a ten-sided dice

6 Draw a spinner to show each of the following:

a probability of landing on a red section $\frac{5}{8}$

b probability of landing on a red section $\frac{1}{4}$

c probability of landing on a red section $\frac{2}{5}$

7 There are 500 tickets in a raffle. One person buys 10 tickets in the raffle.

a What is the probability that they will win?

b What is the probability that they will not win?

8 A sports club runs a raffle with one prize only. The prize is K200 and there are only 200 tickets being sold. Tickets cost K1.50 each.

a Calculate a person's probability of winning if they buy 10 tickets.

b What is the probability that they will not win?

c Calculate a person's probability of winning if they buy 50 tickets.

d What would be the cost of buying that many tickets?

e What would be the probability of the person not winning?

f How much profit will the sports club make if they sell all the tickets?

9 This table shows the membership of three different sports teams.

	Team A	Team B	Team C
Males	5	12	6
Females	8	2	6

a Which team has the largest number of members?

b In which team is there more chance of a team member being a male than a female?

c In which team is there an equal chance that a team member is a male or a female?

Unit two

10 Calculate the volume of the following rectangular prisms:

a a cube with side lengths of 3 m

b width 12 cm, length 8 cm and height 5 cm

c height 5 m and a base area of 20 m^2

d length and width both 12 cm and height 15 cm

11 Calculate the capacity of containers with the following volumes:

a 20 cm^3

b 35 cm^3

c 1300 m^3

d 1.75 m^3

12 A rectangular pool has dimensions 3.2 m $\times$ 4.8 m $\times$ 5 m. How many litres of water will it take to fill the pool?

13 a A petrol container holds 5 gallons. How many litres is that?

b A bucket has a capacity of 25 L. How many quarts will it hold?

c The pond can hold 25 L of water. How many gallons will fill the pond?

14 Order the following measurements from the longest to the shortest.

100 m, 100 feet, 250 mm, 10 miles, 10 inches, 1 m, 2 yards, 10 cm, 10 yards, 1 km, 50 inches

15 Increase:

a 24 by 25%

b 72 by 36%

c 53 by 3%

d 126 by 17%

16 Decrease:

a 65 by 15%

b 56 by 18%

c 84 by 6%

d 108 by 31%

17 Calculate the percentage increase when:

a 40 becomes 75

b 30 becomes 55

c 6 becomes 48

d 27 becomes 54

e 18 becomes 66

f 120 becomes 144

18 A store was marking all items down by 15%. What will be the new price for each of the following items?

a a dress that was K40

b a handbag that was K15

c a jumper that was K25

d a pair of sandals that was K18

19 Find the mean, median and mode for the following sets of numbers.

a 7, 11, 18, 3, 23, 6, 3, 19, 12

b 2, 5, 20, 4, 3, 16, 8, 16

c 1.2, 4.5, 2.06, 7.03, 2.6

20 This column graph shows the exam scores of a student during the year.

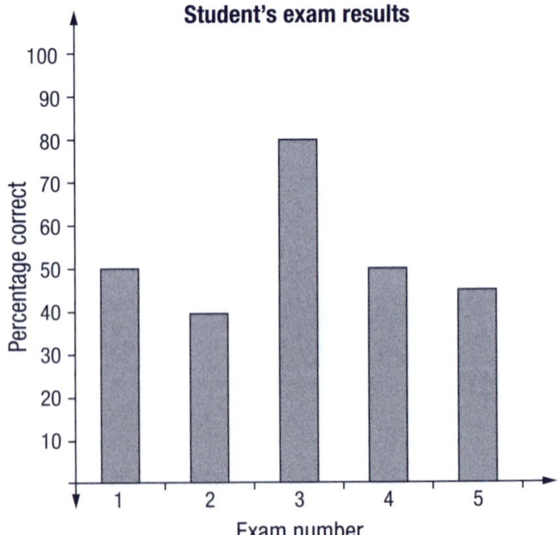

a What does the vertical axis show?

b What does the horizontal axis show?

c What was the difference between the student's best and worst scores?

d Predict the range that you think the student's next exam score is likely to be in.

Graph A: Income from sales ACME ice cream company

Total sales (K): 200, 225, 250, 275, 300, 325, 350, 375, 400

Mon Tue Wed Thu Fri Sat Sun

Day

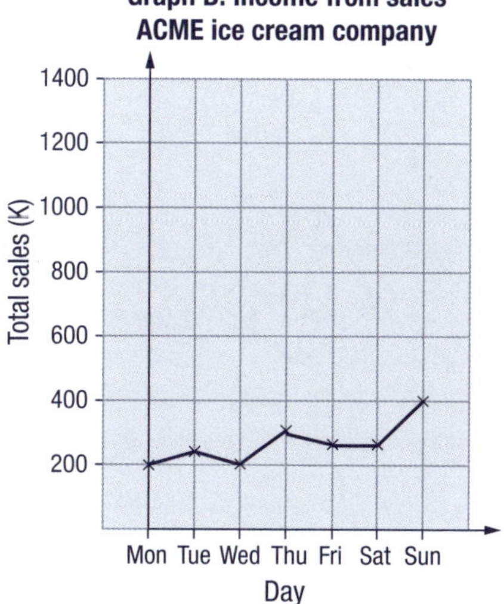

21 The two graphs above show the same information but they appear different.

a Explain how the impression given by the two graphs differs?

b What has been done to create this difference?

c Suggest reasons for using graph A.

d Suggest who might prefer to use graph B. Why?

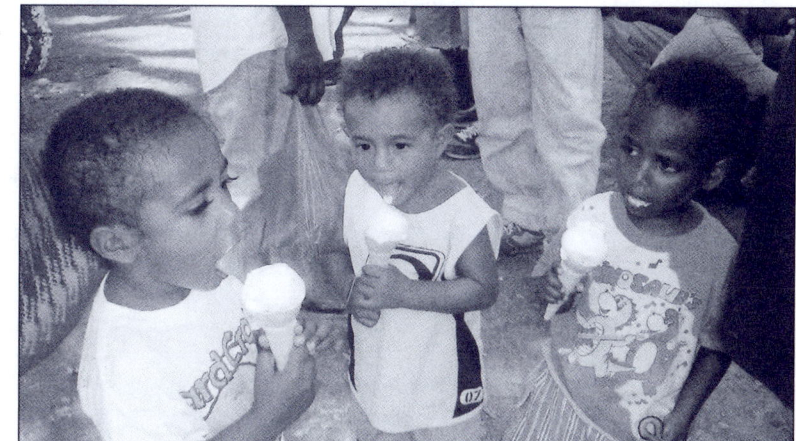

Unit three

22 Draw three different prisms and label each prism with its correct name.

23 Draw three different pyramids and label each pyramid with its correct name.

24 Which of the following nets can be folded to form a prism?

a 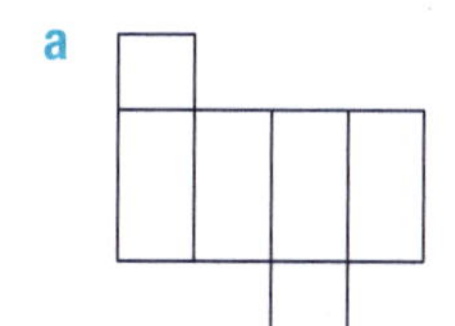

b 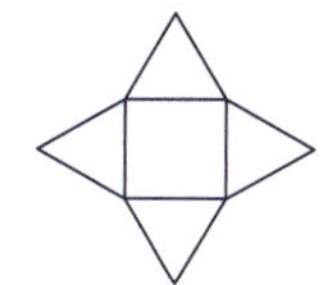

c 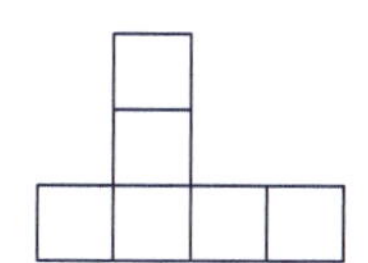

d 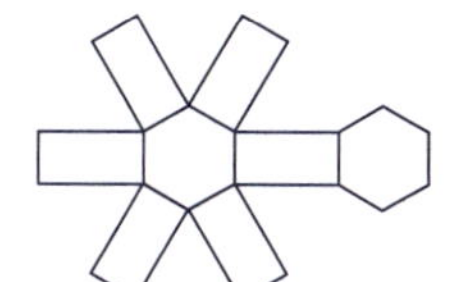

e 

25 Use the correct formula to find the volume of:

a a cube with a side edge of 12 cm

b a cylinder with a base area of 33 cm^2 and a height of 8 cm

c a triangular prism 15 cm high with a base area of 24 cm^2

d

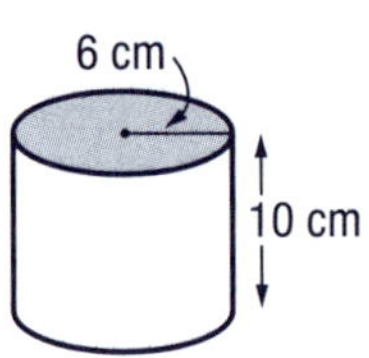

e

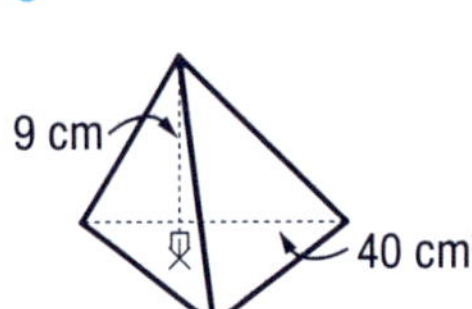

f

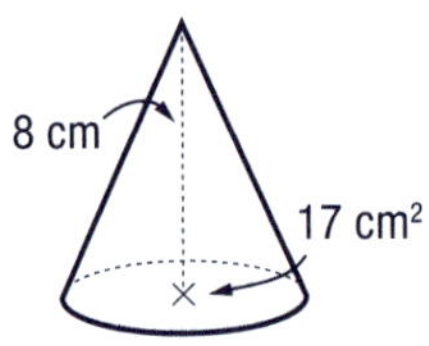

26 Calculate the value of *x* in the following.

a

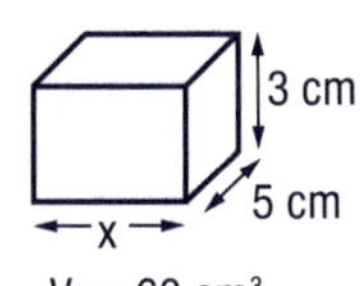

b

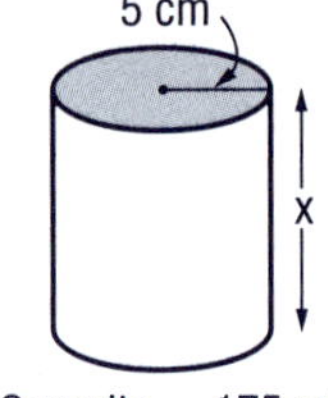

c

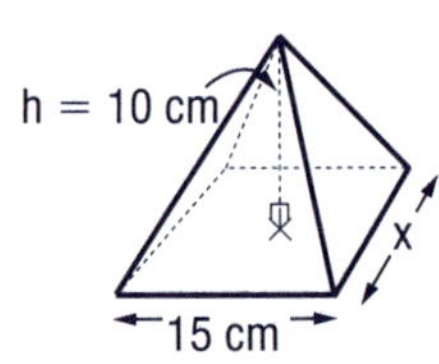

27 Samson has to pack 24 packets each measuring 12 cm long, 4 cm wide and 6 cm high into one large box.

a What are the dimensions of the smallest box that Samson could use?

b Draw and label a plan to show how the packets would fit into the box.

28 A test contains 45 questions.

a How many questions are in the first one-third of the test?

b What fraction of the test has been completed after five questions?

c How many questions will the student still have to complete when they have answered $\frac{3}{5}$ of the questions on the test?

29 Solve the following problems by first converting all the numbers to common fractions or mixed numbers.

a $2\frac{3}{4} + 0.25$

b $2.8 + 3\frac{3}{5}$

c $4\frac{1}{2} - 2.2$

d $3.1 - 2\frac{1}{4}$

e $1.4 \times 3\frac{7}{10}$

f $2\frac{1}{2} \times 2.03$

g $2.56 \div \frac{1}{4}$

h $4\frac{3}{8} \div \frac{1}{2}$

i $4.5 \div 1\frac{3}{4}$

j $3\frac{7}{10} \times 2.4$

30 What fraction is:

a 2 m of 3.2 m

b 1.5 days of a fortnight

c 1.75 L of 4.5 L?

Additional learning unit

31 Calculate:

a 20^2

b $\sqrt{64}$

c $3^2 + 5^2$

d $\sqrt{36}$

e $\sqrt{81} - 2^2$

32 Use a factor tree to find the prime factors of:

a 72

b 200

c 508

d 256

Assessment

Investigation

This assessment task gives you the chance to show what you understand about nets, volume and capacity in a real-life setting. The work you have done in the units 'Water Sports' and 'In the Bag' will help you to complete this task.

Task 1: Practical investigation – group work

A sports club has decided to sell hot chips and cordial at the ground to raise money for new uniforms. The committee has agreed on the size of the scoop and the quantity of cordial per serve, but they are uncertain about what shape and size each container should be. They are trying to decide between two options for both the chips and the cordial.

Chips

1 Twisting some strong paper to make a **cone** or

2 Using some pre-formed **cylinders.**

(The cost and time needed to make either option is identical.)

Cordial

1 A plastic-coated disposable **box** or

2 A plastic disposable **cylindrical glass.**

Work together as a group first to estimate what volume one scoop might hold, then to determine the dimensions of a cone and a cylinder with the same capacity. Compare both containers and decide which one might be most appreciated by customers and why. Plan a presentation to convince your peers about the advantages of specific options.

Your group will be assessed on your:

- ability to work productively as a group
- evidence of reasonable estimations made prior to calculations
- choice of appropriate problem-solving strategies
- correctness of calculations
- presentation of diagrams labelled with appropriate dimensions
- overall effort and persistence
- presentation regarding advantages of specific options compared to others
- ability to complete the task in the time allocated.

Total 40 marks

Assessment

Test

Task 2: Test of basic skills and routine applications

Ratios

1 Simplify the following ratios:

a 54:63 b 49:84 c 88:40 d 300:20

(2 marks)

2 Concrete can be mixed in the ratio 2:4:5, cement, sand and gravel.

a How many buckets of cement should be mixed with 30 buckets of gravel?

b How many buckets of sand should be mixed with 8 buckets of cement?

(2 marks)

3 On a truck there are 8 pigs, 22 chickens and 10 geese. Write each of the following as ratios in their simplest form:

a number of pigs to geese

b number of chickens to total animals on the truck

c number of geese to chickens

d number of chickens and geese to number of pigs

(4 marks)

Probability

4 When a box was tossed, the students identified three ways it could land: on its end, on its side or flat. This table shows the outcomes when the box was tossed 50 times.

Outcome	End	Side	Flat
Number of times	10	5	35

Calculate the Pr that the tossed box will land:

a on its end b flat

(4 marks)

5 A spinning wheel has the numbers 1–50 evenly spaced around it. Only 50 tickets are sold each spin. Calculate the probability of someone winning if they bought:

a two tickets b all the odd numbered tickets between 30 and 50

(4 marks)

Volume

6 Calculate the volume of the following uniform solids:

a a rectangular prism 12 cm long, 3 cm wide and 4 cm tall

b a cylinder 8 cm tall with a radius of 6 cm

(4 marks)

Assessment

Test

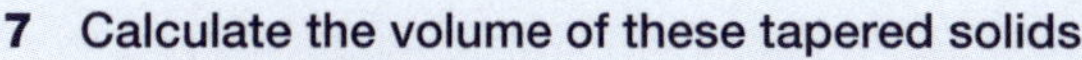

7 Calculate the volume of these tapered solids:

a

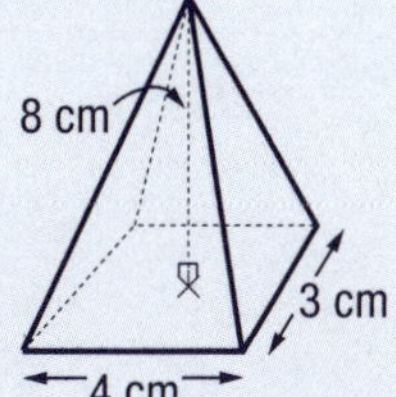

b

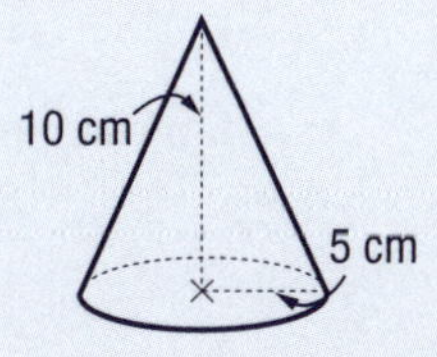

(4 marks)

8 Calculate the capacity of the following containers:

a a cube with a volume of 24 cm^3

b a cylinder 8 cm tall and a diameter of 5 cm

c a cone 12 cm tall with a base area of 15 cm^3 (6 marks)

9 A rectangular spa pool has dimensions 2.5 m × 1.8 m × 1.5 m. How many litres of water will it take to fill the spa?

(2 marks)

Capacity

10 Order from the largest quantity to the smallest quantity.

a 3 quarts, 100 fl. oz, 10 gallons, 5 L

b 1 L, 2 gallons, 30 fl. oz, 5 quarts

c 5 L, 4 quarts, 10 pints, 2 gallons

d 100 fl. oz, 4 gallons, 500 fl. oz, 10 L

(4 marks)

11 Order from longest to shortest.

a 0.3 m, 10 yards, 15 feet, 5 km, 20 cm

b 0.5 km, 2 miles, 300 cm, 12 inches, 1 yard

c 1 mile, 20 yards, 10 m, 3 mm, 0.3 km

d 5 m, 3 km, 50 yards, 200 cm, 10 inches

(4 marks)

Decimals

12 An athlete's practice times for the 100 m sprint were: 13.85 sec, 13.06 sec, 13.69 sec.

a What was her best time?

b By how much did she improve between her best and worst times? (2 marks)

13 Otto's trial times for the 50 m backstroke were 32.74 sec, 31.93 sec, 33.09 sec.

a What was Otto's mean time?

b Assuming his speed was constant, how long did it take Otto to swim one metre in his first trial?

(4 marks)

Assessment

Test

Percentage

14 Calculate the percentage increase or decrease when:

a 80 becomes 95 b 36 becomes 48 c 35 becomes 28 d 85 becomes 70

(4 marks)

Data

15 A student scored these test results during the year: 72.5, 61.5, 70, 73, 62, 70, 74

a What was her median score? b What was her mode score?

(2 marks)

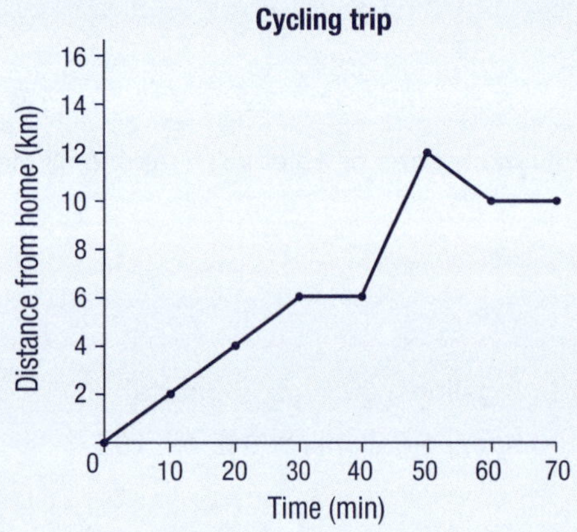

16 The graph shows the distance Rosa cycled one morning.

a Did Rosa start her journey from home? Explain.

b How many times did Rosa stop during her journey?

c At what period was Rosa cycling the fastest? Explain. (3 marks)

Nets

17 Draw the net for:

a a prism b a triangular pyramid c a tapered solid other than a pyramid

(3 marks)

Fractions

18 In a class of 40, there are 24 boys. One-third of the boys and $\frac{3}{8}$ of the girls play tennis. How many tennis players are there in the class?

(2 marks)

Total 60 marks

For Teachers

For Teachers

The Oxford Grade 8 Mathematics Program was developed to reflect the teaching approaches described in the Upper Primary Teachers Guide (2003) produced by the Department of Education, Papua New Guinea. The Teachers Guide provides valuable information about curriculum development, appropriate use of context, selection of suitable learning and teaching strategies and how to plan and carry out effective assessment. The Grade 8 Mathematics Program should be used in conjunction with the Teachers Guide to help teachers implement the Grade 8 Mathematics Syllabus.

The Grade 8 Program comprises four topics across two books. Book A and Book B each contain two topics and it is expected that each topic will take one term to complete.

Each topic is divided into three Learning Units and one Additional Learning, Revision and Assessment unit.

All learning outcomes for each Mathematical Strand have been covered in the Oxford Grade 8 Mathematics Program. On later pages a chart matches the learning outcomes to the various Learning Units.

A key focus of the Grade 8 Mathematics books is that they are based on particular contexts. Four broad topics of general interest have been chosen and each Learning Unit examines a specific context within these broader topics.

The mathematical concepts have been organised to build upon each other as students work through the two books. This means that in order to make learning as effective as possible, students should understand the concepts contained in Book A before beginning Book B. However, depending on your students' needs, you may decide to alter the order in which they complete the topics.

Planning your teaching

In Upper Primary 180 minutes per week has been allocated to mathematics. The National Department of Education's policy is to allow flexibility in timetabling and you may, for example, plan to teach five 36-minute lessons or four 45-minute lessons a week.

As a general guide teachers should aim to finish one topic in the Grade 8 Mathematics book each term. The individual lessons vary in length but each would be expected to take a minimum of 36 minutes to complete and in some cases may take two or three lessons. Plan to spend about 3 weeks on each Learning Unit and one week on the Revision Unit which includes a formal test.

Using contexts

Recent reforms in education have highlighted the need to support students so they can use the mathematics they learn in realistic, practical situations. Using a context for teaching mathematics has the added advantage of helping students to see the meaning of the concepts and of making the posing of mathematical problems easier to understand.

It is important that the curriculum builds links for students between the mathematics that they do at school and its application to real-life situations.

Explain why you are using the contexts

Linking new mathematical concepts to familiar and interesting contexts enables students to integrate their learning in a way that relates more readily to their own lives. The contextual approach requires students to use knowledge and skills from a range of sources to find solutions using the methods and levels of accuracy appropriate to the real-life context.

Discuss the context with students

The context is useful for showing students how the mathematics is used in real-life. Generally, contextualising the mathematics also makes it easier to understand. However, in some cases the context can make the mathematics more difficult to learn. For example, if you want to talk about sea-shells, students from Highlands regions may have difficulty understanding what you mean. It is very important that students understand the context that is being used. Teachers should adapt the language and/or the context to best meet the needs of their students.

Transfer the concept to another context

Mathematical knowledge and skills gained in one context are only useful if they can be applied in some other context. While it is helpful to initially teach a concept using a familiar context, it is important that the student finally be asked to solve a similar problem using a different context. Only then can teachers know that the concept has been learned.

Topic introduction

Each topic in the Grade 8 Mathematics Program begins with a double page spread that provides an overview of the broader context and its relevance to the student. The links between the context and the mathematical concepts for each Learning Unit also appear on this double page spread. Use the topic introduction pages to initiate discussion about the context and to find out what students already know about the real-life situations described and the mathematics involved.

The first lesson in each Learning Unit introduces the specific context of that unit within the broader topic. The introduction brings the context into the realm of the individual student and explores their existing knowledge of the context at a personal level.

Catering for diversity

It is the teacher's responsibility to ensure that all students progress in their learning. The lessons in these books have deliberately used contexts that are sufficiently broad enough to cater for a wide range of student interests and backgrounds.

The books include a Glossary and Help Boxes to scaffold student learning, while the Challenges promote and extend thinking. Similarly the Revision Units provide an opportunity for students to demonstrate their understanding by undertaking a student-centred investigation and completing a formal test.

Teaching and learning strategies

Some of the concepts in the books are quite complex and may require a different teaching style. The Papua New Guinea Mathematics Syllabus (2003) promotes the importance of giving students opportunities in mathematics to work cooperatively, discuss and listen to each other's opinions. It is important that students are clear about the problems they are to solve and that they are provided with adequate time to find solutions themselves without being told how to do them. In some cases it may be appropriate for the class to work together on one question rather than work independently on a series of similar questions.

Language

English is the main language of instruction for Upper Primary, although local vernacular can be used to facilitate understanding and reinforce meaning. The Glossary at the back of each book identifies the key words that students need to learn while working through that book. The first few times students meet these key words in the books, teachers should:

- say the word with the class a number of times
- write the word on the board or on a class list or piece of cardboard
- explain the word using real objects, actions or pictures
- demonstrate how to use the word in a simple mathematical sentence
- ask the students to use the word in a simple mathematical sentence
- tell the students to enter the word in their language vocabulary book or the class dictionary.

Materials

The Grade 8 Mathematics books use a range of different materials. Students learn best when they use real materials to help them explore new concepts. Many of the materials that you will need for the program can easily be collected from the students' environment or made from simple items commonly available. Before students use new materials they should be given time to explore them. This will allow them to concentrate on the task when it is finally set instead of wanting to play with the new materials.

In Grade 8 the intention is for students to move progressively away from using real materials when working with numbers to deal directly with symbols. However, with concepts such as fractions, decimals and algebra it may still be preferable to use concrete materials to assist student understanding.

A list of the main materials needed for each topic appears on the double page spread at the start of each topic. It is important that teachers check this list before beginning the topic and ensure that adequate quantities of the materials are available for the students to use. Some of the mathematical equipment listed may need to be purchased or borrowed from other institutions within the community if the school's resources are insufficient.

Materials and equipment should be stored in strong boxes with lids. Students should be told that they are expected to look after materials and to return them to the appropriate box when they have finished using them. Teachers should check all materials regularly and repair or replace any that are damaged.

Assessment

Assessment is the process of finding out what students know. It is important to assess students in order to evaluate what mathematical concepts they understand, how they learn best, what concepts they are ready to learn and how to most appropriately adjust teaching to further support student learning.

Assessment information should be gathered throughout the year using a variety of assessment methods. Written tests and examinations should only form part of the assessment process. Other valuable information can be gathered during everyday teaching by listening to what students say, observing what they do and looking at samples of student work. It is better to collect a small amount of information at many different times throughout the year than to collect a lot of information a few times during the year.

How to evaluate your students

Assessment should be an ongoing process that begins from the first day at school and continues throughout the year. It is important that teachers do not expect students to have mastered all concepts before they have had the necessary experiences and are genuinely able to demonstrate their use of the concepts successfully.

Using open-ended tasks can provide teachers with meaningful information about students that may not be apparent from student responses to closed tasks. The following example shows the difference between an open and closed task about money.

Closed task: What change should be given if K10 is paid for goods that cost K7.35?

Open-ended task: Describe three different ways to give a customer change from K10 for goods that cost K7.35.

There are many possible answers to the open-ended question. A teacher can learn more about what a student understands about money from their responses to the open-ended question than from their one answer to the closed question. Student responses to open-ended questions can show whether they can:

- look at the problem in different ways
- find more than one possible answer
- explain their strategies
- see patterns and make links between their answers

How to record assessment information

Given the variety of assessment strategies that teachers are encouraged to use, it helps if teachers develop a systematic way for recording information about their students. It is important to keep records up to date and ensure that some data is gathered from each student regularly enough to show changes in their skills and knowledge. Assessment should take place at least fortnightly if not weekly.

Some teachers keep student work sample folders that show evidence of the most recent level of student understanding and achievement. Ideally students should make suggestions and discuss with their teacher about which work samples to include in their folder. As a new piece of work shows improved understanding, the older work sample should be removed from the file. Student work sample folders that are kept up to date are a powerful method of monitoring and assessing student progress.

Another strategy is for teachers to keep notes on a class list. Teachers could choose to observe one to two students in each lesson and make comments about what they observe regarding the students' skills, understandings and motivation. The notes could include both positive or negative information and both typical and unusual events. An example of an annotated class list appears below.

Date	Name	Comments
24 March	Elsie	Added all decimals correctly. Used phrase 'zero point eighty-seven'. (May suggest misconception of decimals as whole numbers instead of parts of whole.)
26 March	Fabian	Confused about placement of decimal point when multiplying decimals by decimals.
26 March	Leti	Completed all decimal additions correctly and asked to show others how to do it.

Tests

These may be short answer or longer exercises. An end of unit test by itself provides information too late to be of value in the classroom. Use the formal test provided in the Revision Unit in conjunction with your on-going forms of assessment. Regular short tests during a unit of work provide timely information about student progress and offer a chance for the teacher to change their teaching approach if required in order to improve student outcomes.

Using assessment information

All assessment information should be used primarily to evaluate student performance as a means of refining the teaching approach. There is no value in collecting assessment information if it is not used to inform teaching for the purpose of improving student performance. If the assessment process highlights gaps in a student's knowledge and understandings, it is a signal for the teacher to try a different approach. Assessment helps the teacher to provide better learning opportunities for the student to ensure the desired outcomes are achieved.

Outcomes Map

Strand	Learning Outcome	Topic	Unit
Number and Application	8.1.1	Gadgets And the Winner is… A Lot Like Me The Tourism Market	In the Tool Shed In the Bag Different People, Different Sizes Growing and Changing Where Do Tourists Go?
	8.1.2	And the Winner is… A Lot Like Me The Tourism Market	Water Sports Growing and Changing Where Do Tourists Go? Travelling Overseas Tourism Comparisons
	8.1.3	Gadgets The Tourism Market	In the Tool Shed Where Do Tourists Go? Tourism Comparisons
	8.1.4	A Lot Like Me The Tourism Market	Growing and Changing Where Do Tourists Go? Travelling Overseas Tourism Comparisons
	8.1.5	Gadgets And the Winner is… A Lot Like Me The Tourism Market	In the Tool Shed The Games People Play Different People, Different Sizes Where Do Tourists Go?
	8.1.6	Gadgets The Tourism Market	Mobile Phones Where Do Tourists Go? Travelling Overseas
	8.1.7	Gadgets	Mobile Phones Additional Learning Unit
	8.1.8	And the Winner is… A Lot Like Me	Additional Learning Unit Additional Learning Unit
Space and Shape	8.2.5	Gadgets	Wheels
	8.2.6	And the Winner is…	In the Bag
	8.2.7	And the Winner is…	Water Sports
	8.2.8	And the Winner is… The Tourism Market	Water Sports Tourism Comparisons
	8.2.9	Gadgets	Wheels
	8.2.10	The Tourism Market	Where Do Tourists Go?
	8.2.11	Gadgets	In the Tool Shed
	8.2.12	Gadgets	In the Tool Shed
	8.2.13	And the Winner is…	In the Bag

Strand	Learning Outcome	Topic	Unit
Space and Shape	8.2.14	The Tourism Market	Where Do Tourists Go? Travelling Overseas
	8.2.16	The Tourism Market	Travelling Overseas
Measurement	8.3.1	A Lot Like Me The Tourism Market	Different People, Different Sizes Travelling Overseas
	8.3.2	A Lot Like Me	Different People, Different Sizes
	8.3.3	A Lot Like Me The Tourism Market	Different People, Different Sizes Travelling Overseas
	8.3.4	Gadgets The Tourism Market	Mobile Phones Travelling Overseas
	8.3.5	Gadgets The Tourism Market	Mobile Phones Where Do Tourists Go? Travelling Overseas
Chance and data	8.4.1	And the Winner is... A Lot Like Me The Tourism Market	Water Sports The Past and the Future Travelling Overseas Tourism Comparisons Additional Learning Unit
	8.4.2	Gadgets A Lot Like Me	Wheels The Past and the Future
	8.4.3	And the Winner is...	The Games People Play
	8.4.4	And the Winner is...	Water Sports
	8.4.5	A Lot Like Me The Tourism Market	Different People, Different Sizes Travelling Overseas Tourism Comparisons
	8.4.6	Gadgets A Lot Like Me	Wheels Different People, Different Sizes
	8.4.7	Gadgets And the Winner is... A Lot Like Me The Tourism Market	In the Tool Shed The Games People Play The Past and the Future Where Do Tourists Go? Travelling Overseas Tourism Comparisons
Patterns and Algebra	8.5.1	And the Winner is...	In the Bag
	8.5.2	And the Winner is... A Lot Like Me	The Games People Play The Past and the Future
	8.5.3	Gadgets A Lot Like Me	Mobile Phones Different People, Different Sizes

Glossary

Glossary

adjacent Next to each other. For example, in this triangle side AB is adjacent to side AC because they have a common vertex (A).

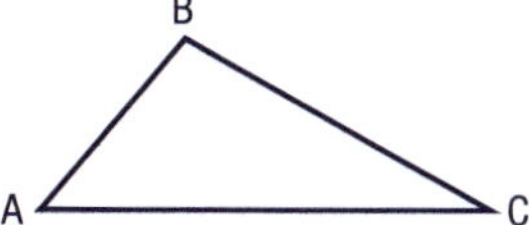

adjacent angles Angles are adjacent if they share a common arm (ray) and a common vertex. For example, in this diagram angle ABC is adjacent to angle CBD.

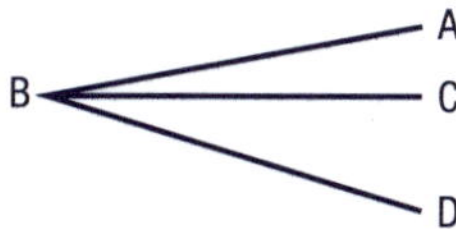

algebra A part of mathematics that uses letters and symbols to stand for unknown values.

alternate angles A pair of equal angles formed when a straight line crosses a pair of parallel lines. The alternate angles are both between the parallel lines and on opposite sides of the transversal. In this diagram *w* and *z* are alternate angles and *x* and *y* are alternate angles.

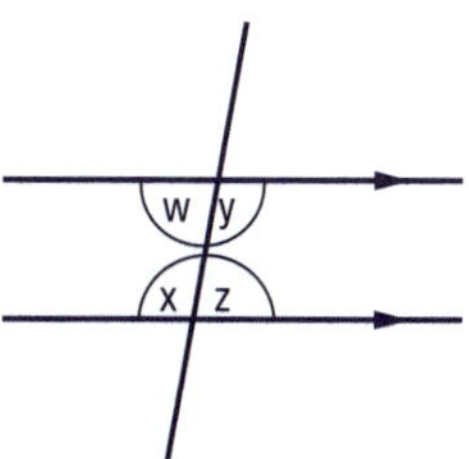

apex The point or vertex that is furthest from the base of a solid or plane shape.

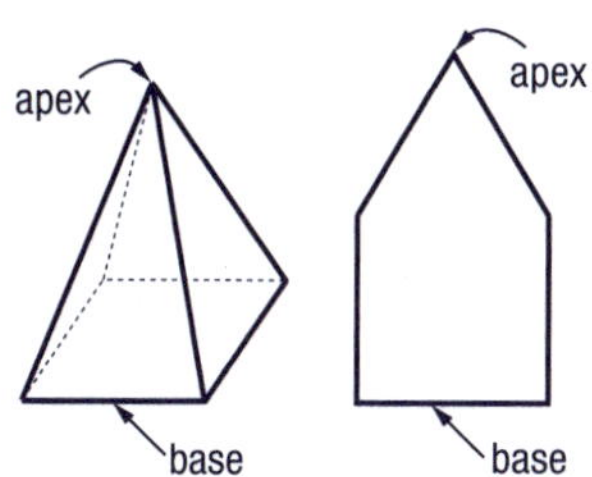

arc A part of a circle or other curved line. You can use a pair of compasses to draw an arc.

area A measure of the total surface of a shape or object. Area is measured in square centimetres (cm^2), square metres (m^2), hectares (ha) or square kilometres (km^2).

calibrate To mark a scale on a measuring instrument. For example, a small ruler is calibrated in centimetres and millimetres; a protractor is calibrated in degrees.

capacity The amount of liquid that a container can hold. This is measured in millilitres (mL), litres (L), kilolitres (kL) and megalitres (ML). There is a close relationship between capacity and volume.

circumference The distance around the edge of a circle.

co-interior angles A pair of angles that are formed when a straight line (called a transversal) crosses a pair of parallel lines. The co-interior angles are on the same side of the transversal and between the parallel lines. Here there are two pairs of co-interior angles: *c* and *d*, and *a* and *b*. The angles in each pair add up to 180°.

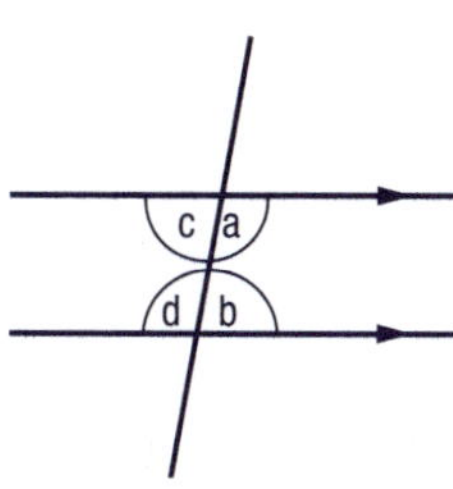

column (or bar) graph A graph that uses horizontal or vertical columns of varying lengths to represent information. The lengths of the columns vary according to the number of items or values that are represented.

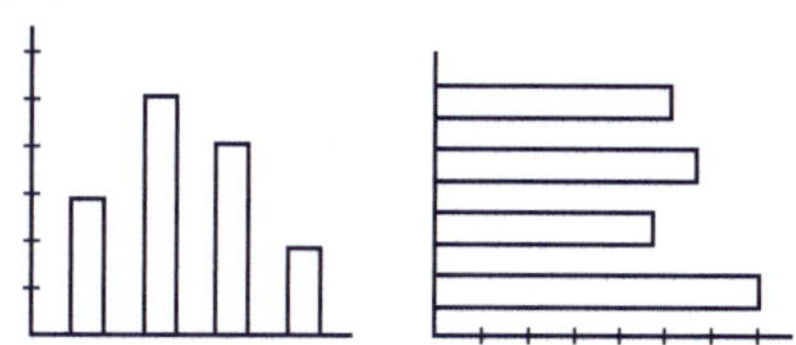

complementary angles Two acute angles that add together to equal 90°.

$\angle ABC + \angle CBD = 90°$

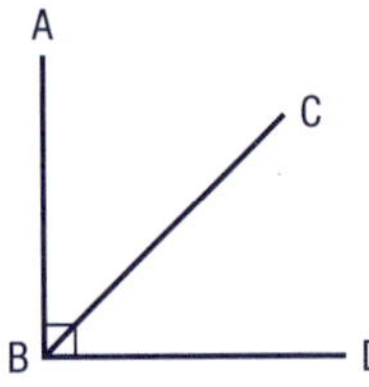

compound shape A shape that is made up of a combination of two or more smaller shapes. The shapes shown here are compound shapes.

concave polygon A straight-sided plane shape that has at least one interior angle that is greater than 180° (a reflex angle). It looks like a vertex has been 'pushed in' towards the inside of the polygon. The polygons shown here are concave and the interior reflex angles are marked.

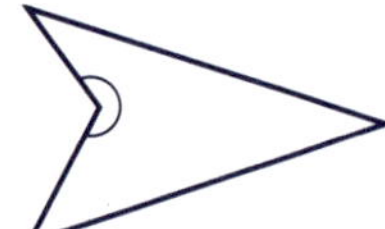

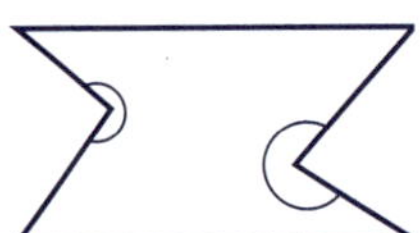

concentric circles Circles that have the same centre point.

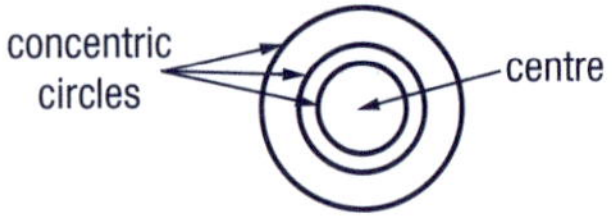

cone A solid shape with a curved surface and one point (vertex).

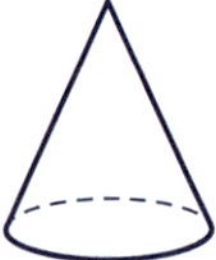

congruent Two shapes are congruent if they have the same size and shape.

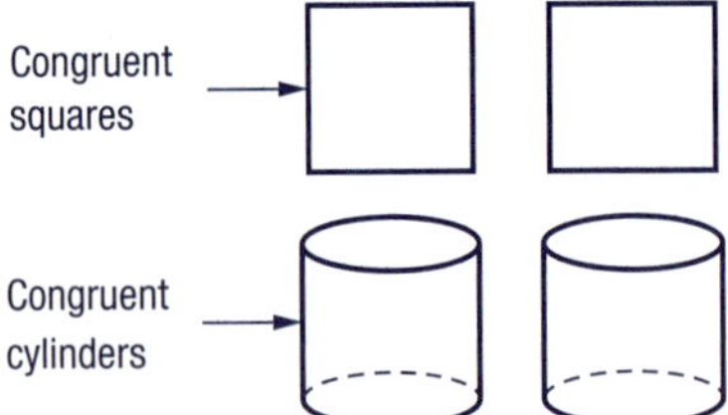

corresponding angles When two lines are crossed by another line (that is called the transversal), the angles in matching corners are called corresponding angles. Corresponding angles are located on the same side of the transversal and are always equal. In this diagram $a = e$, $d = h$, $b = f$ and $c = g$.

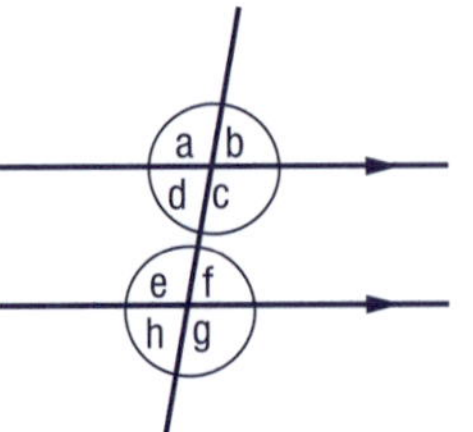

cross-section The face formed when a solid is cut through by a plane. These diagrams show cross-sections of a cylinder and a cube.

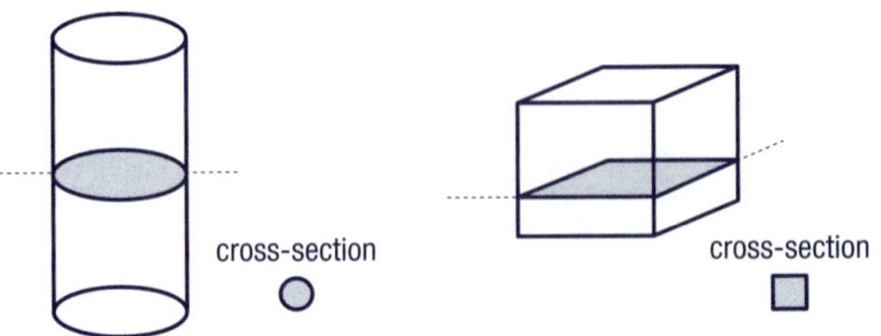

curved surface A surface that is not flat. Many solids have curved surfaces, for example, spheres and cylinders. Cylinders have both curved and flat surfaces.

cylinder A solid with two equal circular faces, one at each end, and a curved surface. Many tins and cans are in the shape of cylinders.

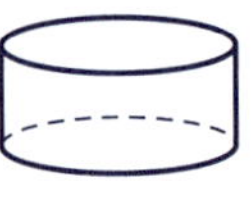

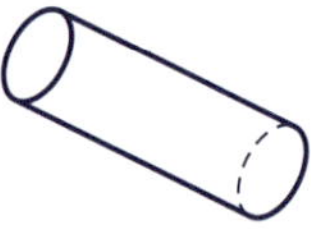

data Also known as *statistics*. A set of data is information (numbers or words) collected as part of a survey or questionnaire. For example: What colour is most popular among students in our grade? The data set is the number of students who like each colour. The set of data can be shown in a table or on a graph to help us interpret it better.

decagon A polygon with ten straight sides. A regular decagon has ten sides of equal length and ten equal angles.

diameter The length of the straight line that passes from one side of a circle to the other through the centre.

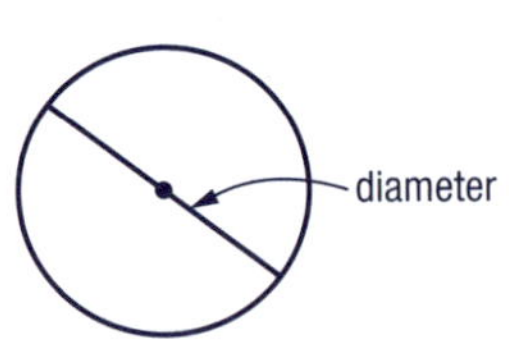

directed number A whole number that has a + or – sign to indicate a positive or negative direction from zero. A directed number is also called an *integer*. We can show them on a number line.

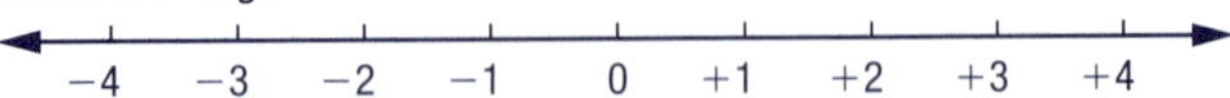

displacement A method for finding the volume of solids, especially those with an irregular shape. The solid is put into a calibrated measure of water and the amount that the water rises on the scale of the measure is the same as the volume of the solid. For example, if a solid displaces 30 millilitres (mL) of water, then its volume is 30 cubic centimetres (cm^3).

distributive law It is one of the basic laws of algebra. The distributive law says that $a(b + c) = ab + ac$. In doing so, it simplifies many multiplication questions by getting rid of the brackets.

dodecahedron A solid that has twelve faces. The faces of a regular dodecahedron are congruent, regular pentagons.

elements of a set A group of numbers, shapes or items with a particular thing in common. For example, the numbers 2, 4, 6, 8, 10, 12, 14, 16 and 18 are the set of even numbers less than 20. The things that belong to a set are called *elements*, so the set described has nine elements.

ellipse A curved shape that looks like an elongated circle.

evaluate To find the value of. If we know the value of a pronumeral we can substitute it with a number and solve the equation. For example, to evaluate the following algebraic expression substitute the pronumeral (x) with 3: $2x + 5$ becomes $2 \times 3 + 5$ which is 11.

expanding (or expansion) A process used in algebra in which the distributive law is used to remove brackets by multiplying each term inside the brackets by the term outside the brackets. This algebraic expression has been expanded by using the distributive law:

$$2(c + d) = 2 \times c + 2 \times d = 2c + 2d$$

experimental probability It is the ratio of the number of times an event occurs to the total number of trials. For example, to find the experimental probability of winning a game, play the game many times then divide the number of games won by the total number of games played.

exterior angle Also called an *external angle*. An angle that is outside a plane figure. It is formed when one of the sides of the shape is extended beyond the vertex.

factor Any number that you can divide into another number without leaving a remainder, for example, factors of 6 are 1, 2, 3 and 6; factors of 10 are 1, 2, 5 and 10. The smallest factor of a number is always 1 and the largest factor is always the number itself.

factorise To simplify an algebraic expression or a whole number by breaking it down into factors. For example,

$$4y + 8 = 2 \times 2 \times y + 2 \times 4 = 2(2 \times y + 4) = 2(2y + 4).$$

formula (*plural*: formulae) A rule that shows how to work out an unknown value by using known values. It is commonly expressed in algebraic symbols, for example, the circumference of a circle is twice its radius times pi so the formula is $C = 2\pi r$.

hexagon A polygon that has six straight sides. A regular hexagon has six sides of equal length and six equal angles.

highest common factor (HCF) The largest number that divides exactly into each number in the group. For example, the highest common factor of 30 and 48 is 6 because the factors of 30 are 1, 2, 3, 5, 6, 10, 15, 30 and the factors of 48 are 1, 2, 3, 4, 6, 8, 12, 16, 24, 48. We can see that the highest factor that is common to both groups of factors is 6.

icosahedron A solid that has twenty faces. The faces of a regular icosahedron are congruent, equilateral triangles.

imperial measurement A system of measurement that was used before the metric system was introduced. Some countries still use a mixture of imperial and metric measures as the conversion to metric requires education of people and changes to machinery and production equipment. The United States of America is the only major country that has retained a non-metric system. Some of the units used in the imperial system are inches, feet, yards, miles, ounces, pounds, pints, quarts and gallons.

index notation A short way of writing large numbers. The index or exponent shows the number of times the base number is multiplied together to give the product. For example: in 5^3, 5 is the base number and 3 is the index. 5^3 is read as '5 to the power of 3' or '5 to the third power' and means $5 \times 5 \times 5$ (125). Index notation is also known as *exponential notation*. The plural of index is *indices*.

inscribed square A square that fits exactly inside a circle so that its corners just touch the circumference of the circle. An inscribed square can be used to help us approximate the area of the circle in which it is inscribed.

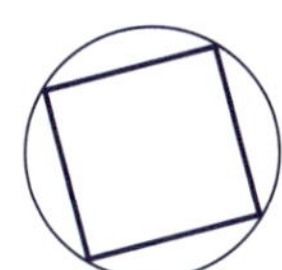

integer A whole number or zero. The integers are:

- positive whole numbers: 1, 2, 3, 4, 5, . . .
- negative whole numbers: –1, –2, –3, –4, –5, . . .
- zero: 0

interior angle Also called an *internal angle*. An angle that is inside a plane figure. It is formed by two adjacent sides of the figure.

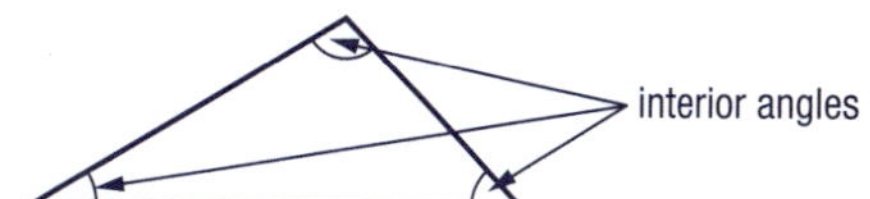

intersecting set It shows the elements that are common to both sets. For example, this Venn diagram shows two sets. The first set is the set of odd numbers to 20. The second set is the set of numbers to 20 that are divisible by 5. The intersection of the sets shows the numbers that are odd and that can be divided by 5. These numbers (5 and 15) are elements common to both sets.

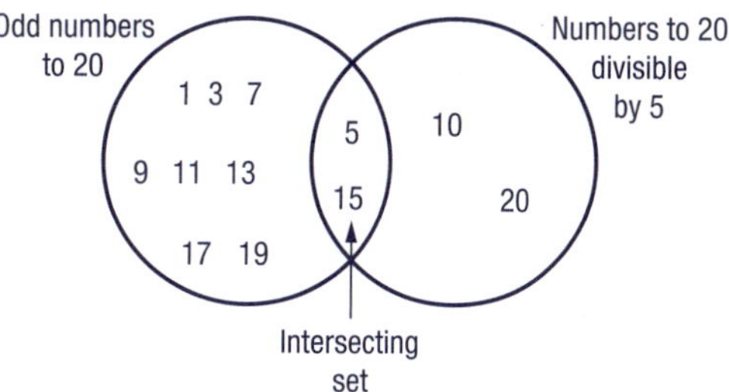

interval The amount of time between two events or happenings. For example, 'The temperature was taken at 10-minute intervals' means that it was taken every ten minutes.

like terms Algebraic terms with the same pronumerals and indices. Like terms can be added and subtracted, whereas terms that are not like cannot be added or subtracted. Examples of like terms include: $5x$ and $2x$, $3x^2$ and $4x^2$, $3xy$ and $-5xy$. Examples of terms that are not like include: $2a$ and $2b$, $3xy$ and y, a^2 and b^3.

line graph A graph that shows a trend or relationship. It shows how two pieces of information are related and how they vary depending on one another. It can be used to show how something changes over time. Segments of straight lines connect points that represent certain data. A line graph has a vertical axis (*y*-axis) and a horizontal axis (*x*-axis).

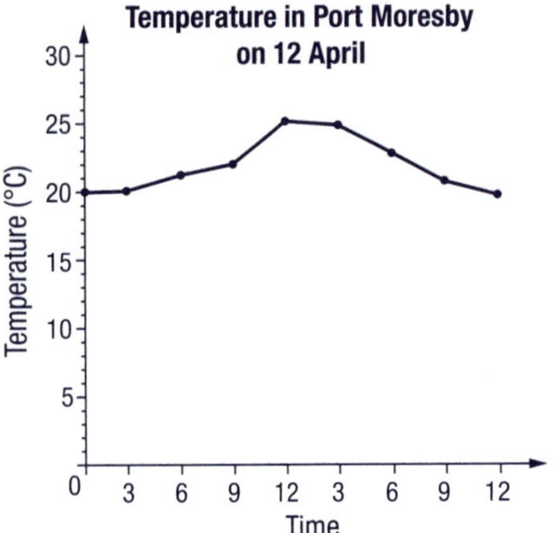

long-run proportion The experimental probability of an event occurring is the proportion of times that it occurs 'in the long run' (over a large number of trials). Experimental probability is calculated using long-run proportion.

lowest common multiple (LCM) The smallest number that is a multiple of both of the given numbers. For example, to find the lowest common multiple of 4 and 6 we first find the multiples of each number. For 4: 4, 8, (12) 16, 20, 24 . . . and for 6: 6, (12) 18, 24, 30, 36 . . . So the lowest common multiple of 4 and 6 is 12.

maximum The highest or largest value. For example, the highest temperature for a certain day is referred to as the maximum temperature for that day.

mean It is also called the *arithmetic mean* or *average*. It is found by adding a set of scores together and dividing by the number of scores. For example, a student threw 6, 5, 7, 4, 3 and 5 goals in six different basketball games. The mean is found by adding the number of goals together, $6 + 5 + 7 + 4 + 3 + 5 = 30$, and then dividing by 6 (the number of games), which is $30 \div 6 = 5$. So, the mean of the goals thrown by the student is 5.

median The middle number in a set of numbers when the numbers are arranged in order. If there is no middle number because the number of scores is even, the mean of the two middle numbers is taken. In the set of numbers 3, 5, 8, (12) 13, 17, 26, the median is 12.

metric system A decimal system of measurement. It is a relatively modern system that was developed in France during the 1790s. It is a simple system that uses multiples of 10 to produce larger and smaller units. Common units used in the metric system are centimetres, metres, kilometres, grams, kilograms, millilitres and litres.

minimum The lowest or smallest value. For example, the lowest temperature for a certain day is referred to as the minimum temperature for that day.

mixed number A combination of a whole number and a fraction, for example, $3\frac{1}{4}$ and $5\frac{1}{2}$.

mode The most common or frequently occurring number in a set of numbers. For example, in this set of numbers 5 is the most common: 2 5 8 5 3 2 5 5 9 5. There can be more than one mode in a set, and if each number only appears once (or the same number of times) then there is no mode.

multiple A multiple of a given number is any number into which it will divide exactly. For example, 4 divides exactly into 4, 8, 12, 16, 20 . . ., so they are all multiples of 4.

negative number A number less than zero. Negative numbers are always written with a minus sign (–) in front of them, for example, -2, -5.8, $-14\frac{1}{2}$.

net A flat shape that will fold up to make a three-dimensional shape.

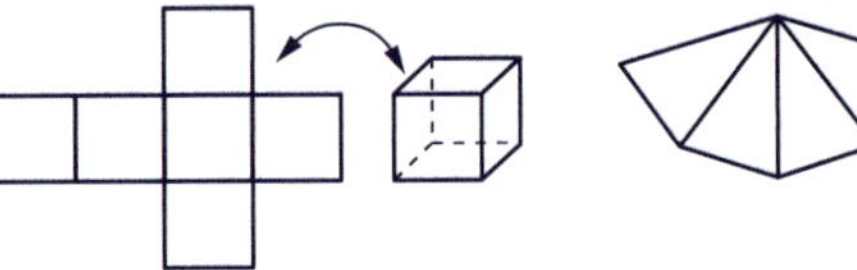

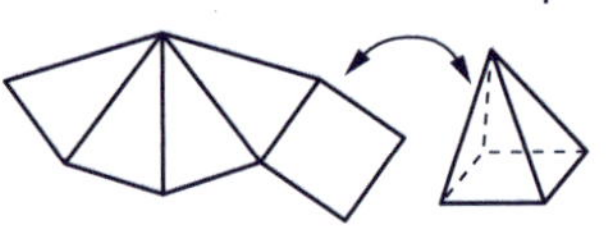

nonagon A polygon that has nine straight sides. A regular nonagon has nine sides of equal length and nine equal angles.

null set Also called an *empty set*. It is a set that does not have any elements. For example, the set of shapes that are both triangles and circles is a null set.

octagon A polygon that has eight straight sides. A regular octagon has eight sides of equal length and eight equal angles.

octahedron A solid that has eight faces. The faces of a regular octahedron are congruent, equilateral triangles.

outcome The result of an experiment or trial.

parallel lines Two or more lines that never meet or cross each other. They always stay the same distance apart. Small arrow marks on lines show that the lines are parallel and lines with the same number of arrow marks are parallel.

pentagon A polygon that has five straight sides. A regular pentagon has five sides of equal length and five equal angles.

percentage A fraction expressed in hundredths. The symbol for percentage is %. For example, $\frac{27}{100} = 27\%$; $\frac{1}{2} = \frac{50}{100} = 50\%$.

percentage change A number can be increased or decreased by a certain percentage. This is referred to as a percentage change. For example, to increase 35 by 25% we say $35 \times 1.25 = 43.75$; to decrease 35 by 25% we say $35 \times 0.75 = 26.25$. To calculate a percentage change we need to know the original number and the increase or decrease. For example, if the original number is 26 and it has been increased to 38 we put the increase (12) over the original number (26) and multiply by 100: $\frac{12}{26} \times 100 = 46\%$, so the percentage change (increase) is 46%.

perimeter The distance around the boundary of an area.

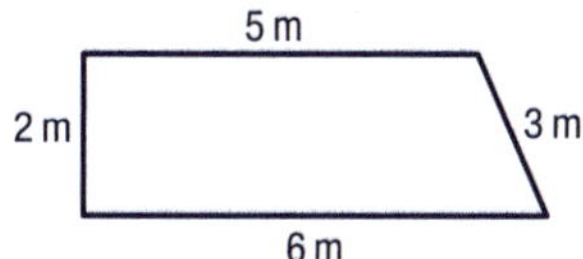

pi The ratio of the circumference of a circle to its diameter. The approximate value of pi is 3.14. Its exact value cannot be worked out. The symbol for pi is π.

pictogram A graph where the information is shown in pictures or symbols. Each picture or symbol represents a certain number or amount of something. This is shown by a key somewhere on or near the graph. A pictogram is also known as a *picture graph* or *pictograph*.

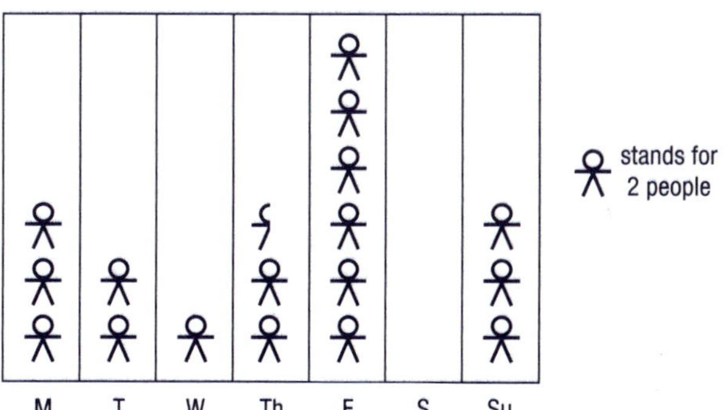

pie chart A graph that is drawn as a circle and shows data as a fraction of the whole, like a pie cut into slices. A pie chart is also known as a *pie graph* or a *circle graph*.

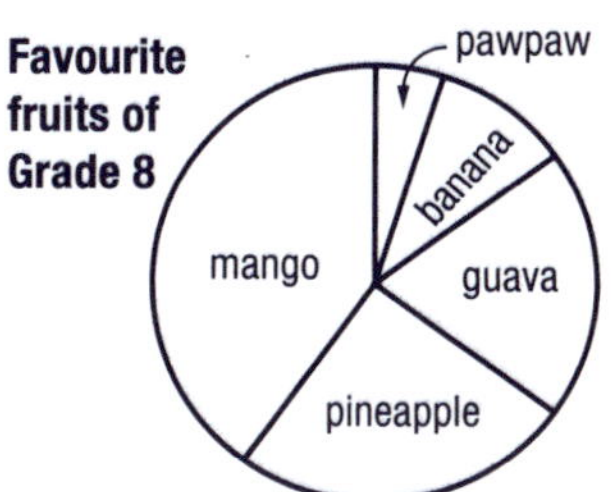

plane surface A flat surface. Many solids have plane, or flat, surfaces.

polygon A flat, or plane, shape with three or more straight sides. Regular polygons are shapes in which all sides and angles are equal, for example:

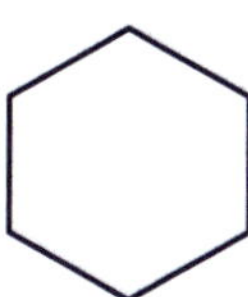

Irregular polygons are shapes in which the sides and angles are not all equal, for example:

prime factor A number that is both a factor and a prime number. For example, the prime factors of 15 are 5 and 3. The prime factors of 12 are 3 and 2.

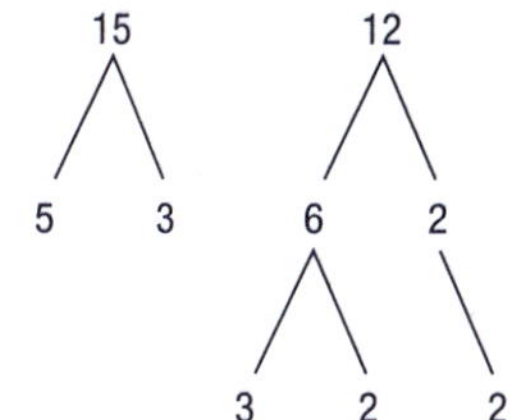

prime number A whole number that has only two factors: itself and 1. That is, it can only be divided by itself and 1 without leaving a remainder. Prime numbers below 20 are 2, 3, 5, 7, 11, 13, 17 and 19. The number 1 is not considered to be a prime number.

prism A solid shape that has two parallel, congruent faces as ends. Prisms are named according to the shape of their ends, for example, a rectangular prism has rectangles on the ends; a square prism has squares on its ends; a triangular prism has triangles on the ends; and a hexagonal prism has hexagons on its ends.

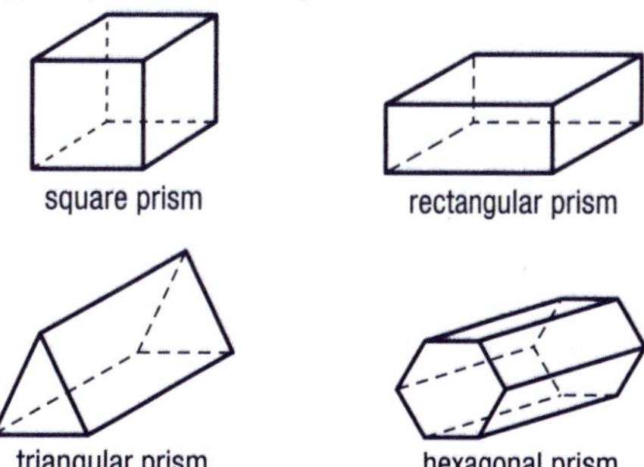

probability The likelihood, or chance, of an event happening. Probability can be expressed using statements such as certain, likely, equal chance, unlikely and impossible. It can also be expressed as a number between 0 and 1 (where 0 means impossible and 1 means certain) and can be written as a fraction, decimal, percentage or ratio.

pronumeral A letter or symbol that stands for an unknown value. For example, in $8 = 5 + y$, the y stands for 3; in $3y - 4 = 17$, the y stands for 7.

protractor An instrument for measuring and drawing angles. Its scale is marked in degrees.

pyramid A solid shape with triangular faces. Pyramids are named according to the shape of their bases. For example:

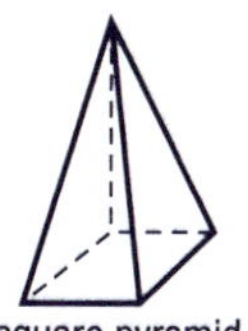
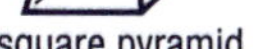

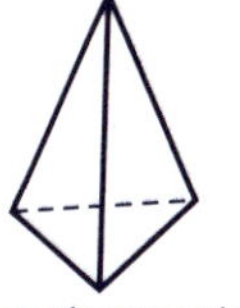

A triangular pyramid is also called a *tetrahedron*.

quadrilateral Any plane shape that has four sides. This includes squares, rectangles, parallelograms, trapeziums, trapezoids and so on. The sum of the angles of a quadrilateral is 360°.

radius (*plural*: **radii**) The length of a straight line from the centre of a circle to its circumference.

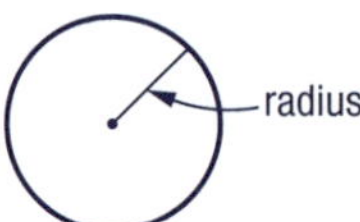

range Also called *spread*. It is the difference between the highest and lowest values of a set of data. For example, in the following set of numbers the range is 15 which is the difference between 20, the highest number, and 5, the lowest number: 5, 7, 10, 11, 14, 15, 17, 18, 20.

rate The comparison between two quantities which may be of different things. For example, the rate of water flow from the water tank was 180 litres per hour. This rate compares volume with time.

ratio A comparison of two quantities, measurements or numbers of items. The symbol : is used to express a ratio. For example, when mixing cordial and water in the ratio 1:4, we mix one part cordial to every four parts water.

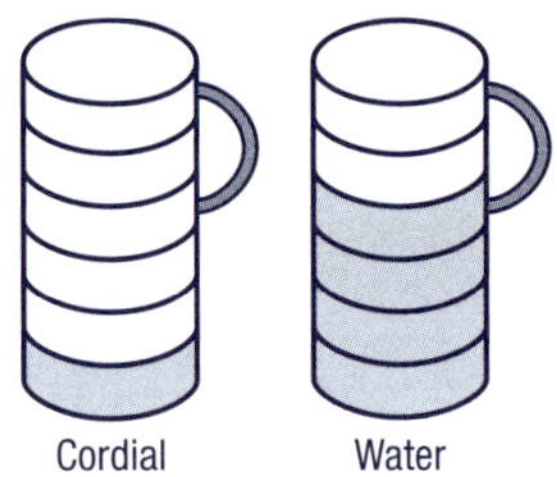

recurring decimal Also known as a *repeating decimal*. It is a decimal fraction with one or more digits that keep repeating. For example, $\frac{1}{3} = 0.3333333\ldots$ is written as $0.\dot{3}$ with a dot over the 3 to show that the 3 is recurring; $\frac{2}{11} = 0.181818\ldots$ is written as $0.\dot{1}\dot{8}$ with dots over the 1 and the 8 to show that 18 is recurring.

right angle An angle that measures 90°. It is quarter of a full turn.

segment Part of a line or shape. A line segment is the part of a line between two points. A circle segment is an area of a circle that is cut off from the rest of the circle and excludes the circle's centre. These diagrams show a line segment and a circle segment.

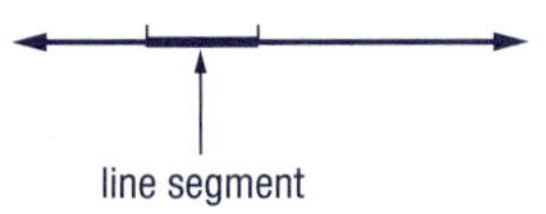

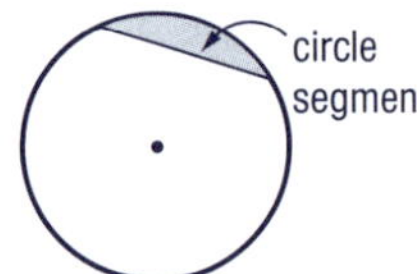

septagon A polygon that has seven straight sides. A regular septagon has seven sides of equal length and seven equal angles.

simplify A process used in algebra to make algebraic expressions less complex. Expressions can be simplified by grouping or combining like terms. For example, $8a + 5b + 3a$ can be simplified to $11a + 5b$; $4x - 8 + 2$ can be simplified to $4x - 6$.

solid A figure with three dimensions. This can be modelled in three forms:

- *solid*: made of solid matter throughout
- *hollow*: made only of the faces, empty inside
- *skeleton*: made only of edges.

sphere A solid shape that looks like a ball. It has one curved surface and no corners or edges.

square number Any number where that number of dots can be arranged in a square shape, for example:

• • (4)
• •

• • • (9)
• • •
• • •

A square number is obtained when a number is multiplied by itself, for example, $2 \times 2 = 2^2 = 4$, $3 \times 3 = 3^2 = 9$, $4 \times 4 = 4^2 = 16$.

square root A number that when multiplied by itself gives the original number. The symbol for square root is $\sqrt{\ }$. For example, 7 is multiplied by itself to give 49 ($7 \times 7 = 49$) so the square root of 49 is 7 ($\sqrt{49} = 7$).

straight angle An angle that measures 180°. It is equal to two right angles or a half turn.

subset A set within a set. All the elements of one set belong to the other set. For example, if Set A includes all the students in your class and Set B includes all the girls in your class, then Set B is a subset of Set A because all the elements of Set B are also in Set A. A subset is defined by a separate boundary.

substitute To replace a letter or a symbol (a pronumeral) with a number. For example, if $y = 5$, the value of $2y + 6$ will be 16.

supplementary angles Two angles that add together to equal 180°. $\angle ABD$ and $\angle CDB = 180°$

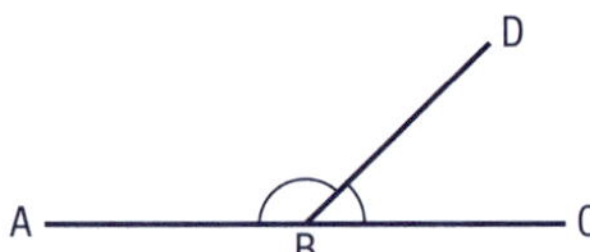

tapered solids Solids such as cones and pyramids that taper to a point. When they are cut parallel to their base the cross-sections are the same shape but smaller. For example:

tetrahedron A solid that has four faces. The faces of a regular tetrahedron are congruent, equilateral triangles. A tetrahedron is also called a *triangular pyramid*.

theoretical probability Probability that can be worked out without an experiment such as when an event has equally likely outcomes. For example, in a throw of a six-sided dice the probability of throwing a 4 is $\frac{1}{6}$. This is because there is one chance in six to throw a 4.

three-dimensional (3-D) When something has the three dimensions of length, width and height it is said to be three-dimensional. Solid shapes are three-dimensional. For example:

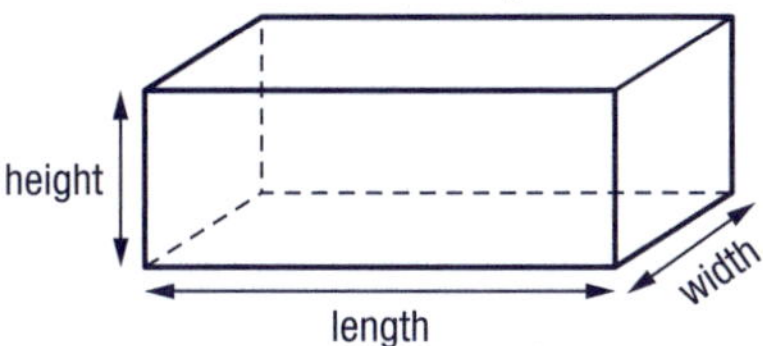

transversal A straight line that cuts across two or more lines. If it crosses a pair of parallel lines it crosses both lines at the same angle.

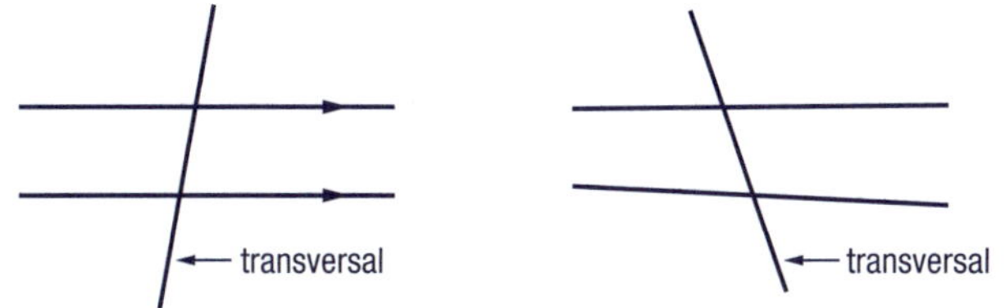

two-dimensional (2-D) When something has the two dimensions of length and width, or length and height, it is said to be two-dimensional. Plane shapes are two-dimensional. For example:

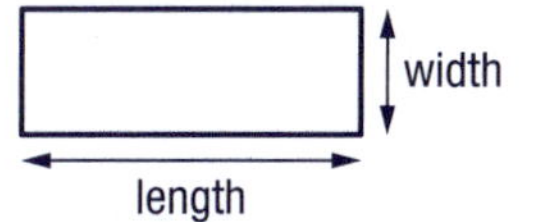

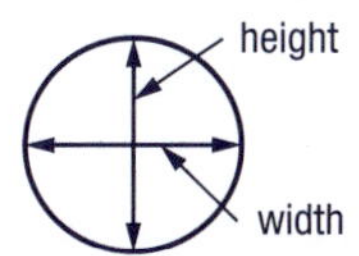

uniform solids Solids that when cut parallel to their ends (bases) have cross-sections that are congruent. Prisms are uniform solids because they are made up of many layers of identical cross-sections. If this triangular prism is cut parallel to its ends anywhere along its length the cross-sections, as shown, will be congruent to the end faces:

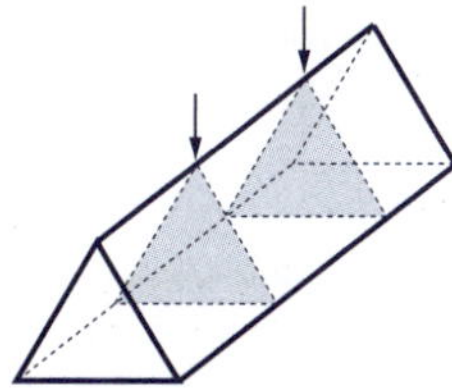

union of two sets This contains all the elements of both sets. For example, if Set A contains odd numbers to 20 and Set B contains the numbers to 20 that are divisible by 5, then the union of the two sets contains the 12 elements of both sets (1, 3, 5, 7, 9, 11, 13, 15, 17, 19, 10, 20), including the two elements that are common to both sets.

variable A pronumeral that can change in value. For example, in the equation $3x - 2 = y$, both x and y are variables.

Venn diagram A way of displaying data when items belong in more than one set. It shows the relationship between sets. The places where the circles intersect show the items that belong in both sets.

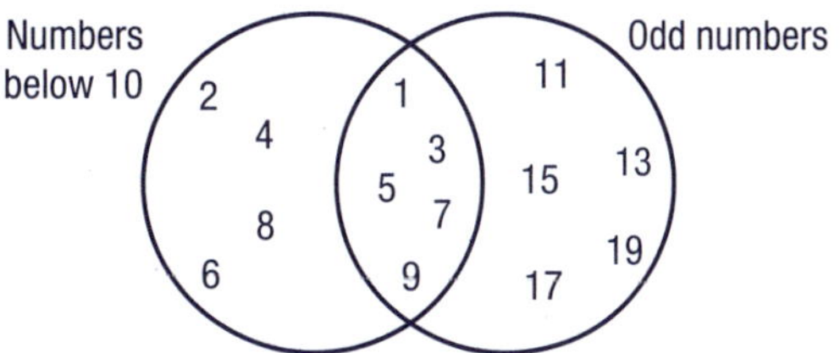

vertex (*plural*: vertices) A point at which two or more lines or edges meet to form an angle or corner.

vertically opposite angles When two lines intersect four angles are formed. The angles that are opposite each other are equal in size. In this diagram *d* and *b* are vertically opposite angles and *a* and *c* are vertically opposite angles.

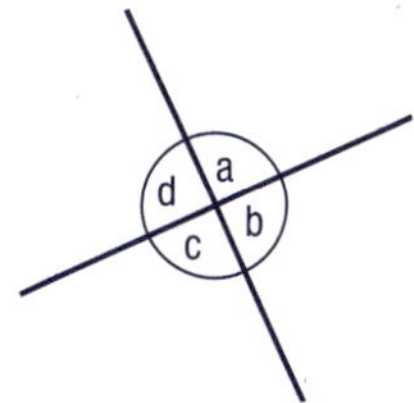

volume The amount of space inside a container or the amount of space a solid occupies. The volume of solids is measured in cubic centimetres (cm^3) or cubic metres (m^3). The volume of liquids is measured in litres and millilitres.

***x*-axis** A horizontal number line used to state the position of a point on a graph. It is at right angles to the *y*-axis and the point of intersection, where the axes meet, is called the *origin*.

***y*-axis** A vertical number line used to state the position of a point on a graph. It is at right angles to the *x*-axis. This diagram shows the relationship between the *x*-axis and the *y*-axis.

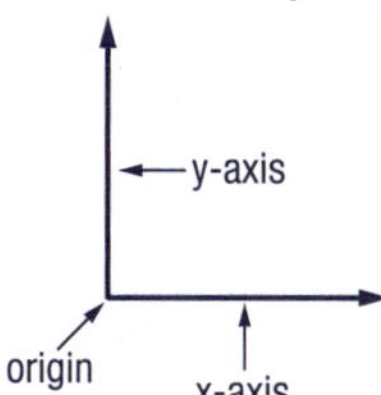

Answers

Topic 1
Learning Unit 1: Mobile Phones

Lesson 1

1–8 Answers will vary.

Lesson 2

1 a 50t per block b 30t per block c 20t per block
d 32t per block e 50t per block f 12t per block

2 a K1.16 b 87t c K2.61
d 58t e 58t f K1.45

3 a 1 min 01 s b 3 min 01 s c 1 min 31 s
d 2 min 31 s e 4 min 31 s f 5 min 01 sec

4 a K1 b K1.20 c K2 d K6 e K4.80

CHALLENGE: K10.20

5 a

	8 a.m.	noon	6 p.m.	10 p.m.
Bernadette	90t	K3	90t	90t
Harold	K1.20	K2.40	K1.20	K1.20
Fegsley	K1.80	K1.80	K1.80	K1.80

b Answers will vary. c Fegsley d Fegsley

6 a 300 seconds
b 990 seconds
c Peak = 360 seconds, Off-peak = 750 seconds
d Peak = 500 seconds, Off-peak = 500 seconds

7 80 minutes

8 a K252 b Answers will vary.

9 a 9t b 90t
c 6 minutes 40 seconds d 161
e 2 minutes 20 seconds

CHALLENGE: Answers will vary.

Lesson 3

1 a 69 toea b K1.20 c K1.05 d 3w toea

2 a 75 toea b K1.05 c 5w toea d 5k toea

3 a Alpha = 75 toea, Beta = 93 toea, Gamma = 87 toea, Delta = 3m toea

b Alpha = 13 toea, Beta = 9 toea, Gamma = 10 toea, Delta = p toea

c Alpha = 125 toea, Beta = 155 toea, Gamma = 145 toea, Delta = 5m toea

d Alpha = 26 toea, Beta = 18 toea, Gamma = 20 toea, Delta = 2p toea

4 K1.55

5

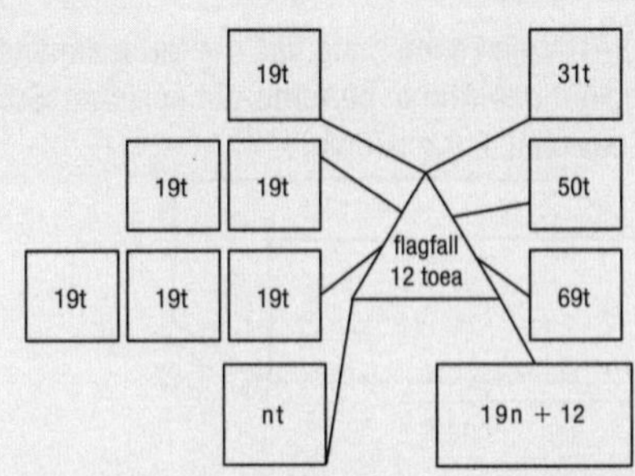

6

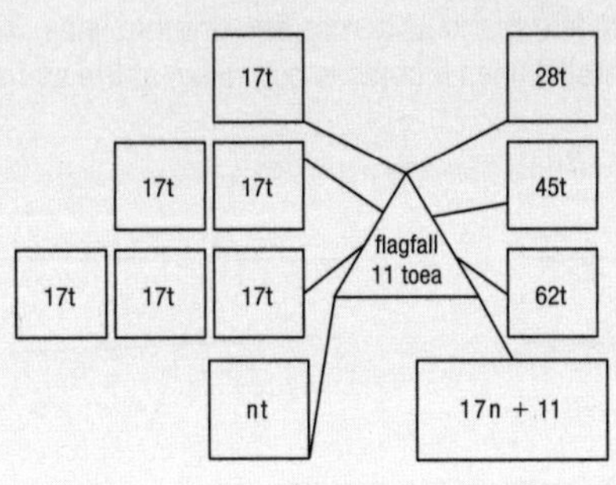

7 a 25 b 70 c 55 d 100

8 a $19n + 12$ b $24n + 12$ c $21n + 30$ d $21n + 15$

9 Wan: $25n + 10$, Tu: $31n + 8$, Tri: $29n + 5$

10 Wan: $13z + 10$, Tu: $9z + 8$, Tri: $10z + 5$

Lesson 4

1

Item	Total times listed	Charge per unit	Total charge
standard calls	3 + 6 = 9 blocks	25 toea per block	9 × 25 = 225 toea (or K2.25)
international roving calls	3 + 4 = 7 blocks	K1.80 per block	7 × 1.80 = K12.60
information calls	1 + 1 = 2 calls	90 toea per call	2 × 90 = 180 toea (or K1.80)
flagfall	1 + 1 = 2 flagfalls	10 toea per flagfall	2 × 10 = 20 toea
wallpaper	1 month of wallpaper	K1.50 per month	1 × 1.50 = K1.50
ringtone	1 ringtone	50 toea per ringtone	1 × 50 = 50 toea
video download	1 video download	K1.20 per download	1 × 1.20 = K1.20

2 a $2x$ and $4x$ b $8ab$ and $2ba$ c x and $5x$
d $3a$ and $2a$ e $2mn$ and mn f $4abc$ and $3abc$

3 a $2p$ b $4a$ c $3g$ d $6a$ e $5x$ f $7y$
g $8k$ h $4n$ i $5m$ j x k k l 0

4 $3s + 14k$

CHALLENGE: Answers will vary.

5 $x - m = y$

6 a $3a$ b $7x$ c $3ab$ d $5mn$
e $7y - x$ f $3p + 2a$ g $2 + 5h$ h $7 - 2k$
i $8h - b$ j $2x + 2y$ k $m + 7mn$ l $3xy + 5x$

7 a $11mn + 6a$ b $8xy - 3x$
c $3ab + 5a$ d $4 + 2k + 4pk$
e $4xy + 3 + 6ab + 2b$ f $6g + 2mn + 3p + 8$

8 a

+	m	$2m$	p	$2p$
m	$2m$	$3m$	$m + p$	$m + 2p$
$3m$	$4m$	$5m$	$3m + p$	$3m + 2p$
p	$p + m$	$p + 2m$	$2p$	$3p$
$2p$	$2p + m$	$2p + 2m$	$3p$	$4p$

b

+	$2x$	$x + 2$	y	$2y + 2x$
x	$3x$	$2x + 2$	$x + y$	$3x + 2y$
$3x - y$	$5x - y$	$4x - y + 2$	$3x$	$5x + y$
y	$y + 2x$	$y + x + 2$	$2y$	$3y + 2x$
$y + 2$	$y + 2 + 2x$	$y + 4 + x$	$2y + 2$	$3y + 2 + 2x$

9 a $3n$ b $4n$ c $7n$

CHALLENGE: Answers will vary.

Lesson 5

1 a 12 b 30 c 17 d 50 e 42 f 45

2 a $x + 25$ b $x - 23$ c $3x$ d $\frac{x}{2}$ e $2x$

3 a $5b + 3c$ b $\frac{9m}{5}$ c $4m + \frac{g}{5}$
d $\frac{h}{3} - \frac{b}{3}$ e $\frac{2g}{3k}$ f $\frac{c}{3} - 2a$

4 a Feb. b 7
c Jan, April, June, July d Feb, March

5

Month	Jan.	Feb.	March	April	May	June	July
Total Kina	18	25	20	17	21	14	9

6

Month	Jan.	Feb.	March	April	May	June	July
Total Kina	21	28	23	20	24	17	10.5

7 a $2p$
b That it charges three times what the Phone-Home company charges.
c $p + s$
d $p - x + s$

CHALLENGE: $s + yp - yx$; where y is an integer bigger than 1.
Note: Any pronumeral not already used can replace y.

Lesson 6

1 a $1 \times (19 + 10) = 29$ b $1 \times (18 + 5) = 23$

2 a $2 \times (19 + 10) = 58$ b $2 \times (18 + 5) = 46$
c $3 \times (15 + 8) = 69$ d $5 \times (12 + 9) = 105$
e $4 \times (15 + 6) = 84$

3 In answers a–h students should show working where they solve the equation in brackets first and then also show working using the distributive law.
a 30 b 20 c 56 d 54
e 39 f 102 g 30 h 60

4 In answers a–e students should show distributive law working.
a $3 \times (19 + 10) = 87$ b $2 \times (24 + 3) = 54$
c $6 \times (10 + 4) = 84$ d $5 \times (11 + 5) = 80$
e $4 \times (9 + 7) = 64$

5 a $5x + 5y$ b $2s + 2t$ c $7a + 7b$ d $4g + 4h$
e $2p + 10$ f $4c + 28$ g $9a + 9$ h $18 + 3m$
i $12 + 6q$ j $24 + 8r$ k $21 + 3e$ l $5g + 15$
m $2t + 12$ n $2j + 8$ o $30 + 6d$ p $9g + 18$

6 a $4p - 4y$ b $5m - 5g$ c $3q - 3p$ d $4t - 4p$
e $8w - 32$ f $2f - 14$ g $5x - 25$ h $25 - 5y$
i $16 - 4j$ j $24 - 6e$ k $32 - 8g$ l $9k - 45$
m $7h - 35$ n $4n - 12$ o $2 - 2r$ p $3t - 21$

7 a $mp + mh$ b $jm + jn$ c $kr + kp$ d $ht + hp$
e $xy - 5x$ f $kt - 4k$ g $ay - 4a$ h $8x - xy$

8 a $8p + 4h$ b $6x + 3r$ c $3g + 9e$ d $5y + 20b$
e $6t - 8$ f $10k - 14$ g $20 - 12d$ h $9g - 18m$

9 a $3m + 2bm$ b $2x - 3yx$ c $2ab + 5b$ d $3kn - 4n$
e $4x - 5yx$ f $4km - 3m$ g $6ps - s$ h $5g + 2hg$

CHALLENGE: $8x^2 - 2x$

Lesson 7

1 a $5(a + b)$ b $3(n + 4)$ c $4(12 - x)$
d $7(6 + y)$ e $\frac{(3 + y)}{2}$ f $\frac{k}{m} + 4$

2 Answers will vary.

3 a $m + n$ b $20m + 20n$

4 a $5b$ b $21a$

CHALLENGE: Answers will vary.

5 $3x + 3y$

6 a That Alice and Edna sent the same number of TXT messages that week.
b That Edna sent double the number of TXT messages that Alice sent that week.
c That Alice sent 5 more TXT messages than Edna that week.
d That Alice sent double the amount plus 3 more TXT messages than Edna that week.
e That Edna sent 6 less TXT messages than Alice that week.
f That Edna sent 2 less TXT messages than Alice that week.
g That Alice sent 3 less TXT messages than Edna that week.
h That Alice sent double the number of TXT messages that Edna sent that week.

7 a $3a + 19$ b $4x - 10$ c $2k + 2$
d $7m - 36$ e $10b + 11$ f $6p - 32$
g $4b + 26$ h $-3 + 2h$ i $3x + 1$
j $8a + 31$ k $16x + 12$ l $2a + 11 - 2b$

8 a $12 + 3l$ b $21 + 11p$ c $13k - 6km$
d $16z - 8zh$ e $29s - 30$ f $19 + 2r$
g $5 + 7h - ch - 2c$ h $6m - 12c + 12$ i $-3k + 63$

Lesson 8

1 a 8 b 4 c 11 d 5 e 14 f 22

2 a 24 b 20 c 2 d 10 e 1.5
f 3 g $\frac{1}{3}$ h 28 i 4 j 20

3 a 19 b 30 c 8 d 14
e 15 f 30 g 3 h 6

4 a 25 b 5 c 20 d 4 e 24
f 21 g 44 h 36 i 10 j 8

5 a 5 b 8 c 38 d 21
e −6 f 36 g 480 h 72
i 17 j 6 k 17 l 72

6 a

x	y
3	10
11	18
19	26
31	38
44	51

b

c	h
10	43
7	31
2	11
1.5	9
0.2	3.8

c

d	t
8	4
14	7
26	13
3	1.5
0.16	0.08

7 $3p + 1 = q$

Lesson 9

1 a 1, 3, 9
b 1, 3, 5, 15
c 1, 2, 4, 5, 10, 20
d 1, 2, 3, 4, 6, 12
e 1, 2, 3, 4, 6, 9, 12, 18, 36
f 1, 2, 4, 13, 26, 52
g 1, 2, 4, 5, 8, 10, 16, 20, 40, 80
h 1, 3, 23, 69
i 1, 2, 3, 4, 5, 6, 8, 10, 12, 15, 16, 20, 24, 30, 40, 48, 60, 80, 120, 240
j 1, 97

2 a 1, 2, 3, 6, 9, 18
b 1, 2, 3, 6, 7, 14, 21, 42
c 1, 2, 3, 6
d 6

3 a 5 b 4 c 16 d 5 e 14 f 3
g 2 h 8 i 6 j 11 k 3 l 7

4 a–d Answers will vary

5 a $2(g + 6)$ b $4(t + 4)$ c $3(d + 5)$ d $4(2w + 1)$
e $3(2s - 1)$ f $12(2k - 1)$ g $5(7x + 3)$ h $9(2y - 5)$
i $22(5 - b)$ j $3(h + 13)$ k $6(7t - 2)$ l $5(5p + 16)$

6 a $5(a + b)$ b $3(7h + 13d)$ c $3(8x - y)$ d $14(2w - g)$
e $5(3t + 5g)$ f $2(-mp)$ g $12(5kn)$ h $7(mn + 2m)$
i $2(3bt + b)$ j $7(3gw-2g)$ k $2(3wv + w)$ l $5(18vr + r)$

7 Answers a–e should show a breakdown of factors in working.
a 684 b 368 c 1696 d 3060 e 4356

CHALLENGE: Answers a–c should show a breakdown of factors in working.
a 43 b 14 c 64

Lesson 10

1 a K2 b −K3 c −K8 d −K4

2 a −K5 b −K5 c −K4 d K0 e K7 f −K5

3 a K8 + K5 − K7 = K6
b −K15 + K11 − K8 = −K12
c K25 − K12 + K5 − K22 = −K4

4 a 1 b 4 c −13 d −11
e 5 f −5 g −1 h −5
i −8 j −4 k −12 l −8
m −21 n −7 o −13 p −53

5 a It decreases by 1 each time.
b They decrease by 1 each time.
c $3 + (-5) = -2$
$3 + (-6) = -3$
$3 + (-7) = -4$

6 a It decreases by 1 each time.
b They increase by 1 each time.
c $3 - (+5) = -2$
$3 - (-3) = 6$

7 a $6 + -5 = 1$ b $13 + +6 = 19$
c $-8 + -4 = -12$ d $-42 + +6 = -36$
e $-12 + -7 = -19$

8 a 2 b −11 c 5 d −3 e −2
f 11 g −7 h −10 i −2 j 3

9 Students should have copied a summary into their books.

10 a $13 + 15 = 28$ b $2 - 18 = -16$
c $52 - 9 = 43$ d $23 + 16 = 39$
e $-12 + 5 = -7$ f $-32 - 5 = -37$
g $-13 - 23 = -36$ h $-9 + 21 = 12$
i $-12 - 13 = -25$ j $-11 + 9 = -2$

11 For questions a–j the equation should be written in the students' books along with either True or False.
a True b True c False d False e False
f True g True h True i True j False

Learning Unit 2: Wheels

Lesson 1

1–6 Answers will vary.

Lesson 2

1 Answers will vary.

2 a Answers will vary.
b Answers will vary (but the answer should be half of the answer to question 2a).

3 Multiply it by 2.

4 The answers to 'circumference' will vary; but should be approximately the same as written below.

Circle	A	B	C	D	E
Radius (cm)	0.5 cm	1 cm	1.5 cm	2 cm	2.5 cm
Diameter (cm)	1 cm	2 cm	3 cm	4 cm	5 cm
Circumference (cm)	3.14 cm	6.28 cm	9.42 cm	12.57 cm	15.71 cm

5 a Answers will vary. b Answers will vary.

6 a 88 cm b 550 mm c 1.5 m d 5.6 m

7 a 200 mm b 19 cm c 1.25 m d 2.3 m

8 a–d Estimates will vary.

9 a–d Estimates will vary.

10 a Students to draw a circle.
b Students to cut out the circle and label it.
c Students to write the measurements on the appropriate part of the circle.

11 Students to mark an arc of 20 cm on the circle and label it.

12 Students to colour and label a segment of the circle.

Lesson 3

1 a 47.5 cm b Estimates will vary.

2 a Estimates will vary.
b Answers will vary (but should be half of 2a).

3 a 60 cm b Estimates will vary.

CHALLENGE: It doubles.
It enlarges five times.

4 a 3.14 m b 62.8 cm c 157 cm d 7.85 m

5 56 520 cm (565.2 m)

6 314 m

7 125.6 m

8 423.9 m

9 a 18.84 m b 219.8 cm c 9.42 m d 471 cm

10 94 200 cm (942 m)

CHALLENGE: a 17.5 m
b 109.9 m

Lesson 4

1 $3.14 \times 80 = 251.2$ cm; hence B

2 $2 \times 3.14 \times 31.85 = 2$ m; hence A

3 $2 \times 3.14 \times 40 = 3.14 \times 80$; hence C

4 $3.14 \times 5 = 15.7$ cm, $3.14 \times 20 = 62.8$ cm, $3.14 \times 3 = 9.42$ cm; hence B

5 a–d Estimates will vary.

6 a 282.6 mm b 87.92 cm c 9.42 m d 3.77 m

7 a–d Estimates will vary.

8 a 6.69 cm b 1.34 m c 23.89 cm d 0.48 m

9 a 4 cm b 1 m c 1.2 m d 19 mm

10 a 6.48 cm b 16 m c 300 mm d 42 m

11 a 131.88 m b 94.2 m c 62.8 cm d 251.2 km
e 157 mm f 172.7 cm g 113.04 m h 628 mm

12 a 62.8 cm b 94.2 mm c 31.4 cm
d 113.04 cm e 62.8 cm f 26.38 km

Lesson 5

1 Student to draw a scale diagram.

2 Estimates will vary.

3 Answers will vary.

4 Students to draw labelled rectangular shapes of the same area as the circles.

5 Students to have a discussion.

CHALLENGE: Yes. Explanations will vary.

Lesson 6

1 a Students to follow the steps to estimate the area of a radius 4 cm circle using an inscribed triangle.
b The estimated area of a circle with a radius of 4 cm, using the inscribed triangle method, is a little more than 32 cm^2.

2 A little more than 72 cm^2.

3 Answers a–d should have working accompanying the answers.
a 50 cm^2 b 800 cm^2 c 18 m^3 d 18 m^3

4 a Students to follow the steps to estimate the area of a radius 8 cm circle using an inscribed square.
b The estimated area of a circle with a radius of 8 cm, using the inscribed square method, is a little more than 128 cm^2.

5 Answers a–d should have working accompanying the answers.
a 98 cm^2 b 242 cm^2 c 128 cm^2 d 20 000 m^2

Lesson 7

1 Answers will vary.

2 200.96 cm^2

3 a Answers will vary.
b It will become more and more accurate (explanations will vary).

4 Answers will vary.

5 a 12.56 cm^2 b 78.5 cm^2 c 113.04 cm^2

6 a 314 cm^2 b 1256 cm^2 c 7850 cm^2 d 31 400 cm^2

CHALLENGE: 78.5 cm^2

Lesson 8

1 a 78.5 m^2 b 153.86 cm^2
c 28.26 m^2 d 2461.76 m^2
e 907.46 m^2 f 706.5 mm^2
g 12.56 cm^2 h 2461.76 cm^2
i 1384.74 mm^2 j 4298.66 cm^2

2 a 346.19 cm^2 b 19.23 m^2 c 39.25 m^2 d 50.24 mm^2
e 38.47 mm^2 f 56.52 cm^2 g 113.04 m^2 h 173.09 m^2

CHALLENGE: 243.49 cm^2

Lesson 9

1 a 22 cm b 33.23 cm^2

2 a 13.14 m^2 b 216.93 cm^2 c 18.13 m^2
d 41.79 cm^2 e 98.24 m^2 f 2575.44 cm^2

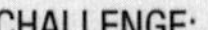

3 a 105.98 m^2 b 200.96 m^2 c 75.36 m^2 d 115.40 m^2
e 188.4 m^2 f 84.78 m^2 g 141.3 m^2 h 7.07 m^2

4 a 115.74 cm^2 b 321.5 cm^2 c 118.63 m^2
d 664.5 m^2 e 674.24 cm^2 f 1793.5 cm^2

CHALLENGE: 1.07 m^2

Lesson 10

1 3 sides, inside angles total 180°, straight sides, more than 2 vertices, 4 sides, inside angles total 360°

2 4 sides, inside angles total 360°, straight sides, more than 2 vertices

3 Straight sides, more than 2 vertices

4 Answers will vary.

5 Answers will vary.

6 a Answers will vary.
b 11
c Students to draw the 3 shapes in the intersecting set.

7 a 8 b 3 c 1 d 12 m^2, 15 m^2, 20 m^2

8

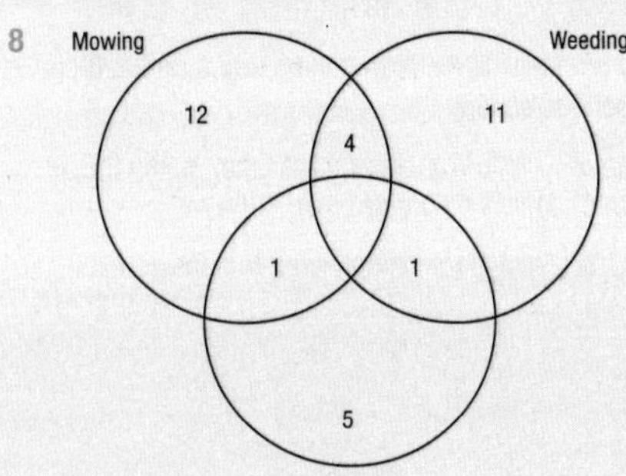

a 12 b 11 c 4 d 18

9

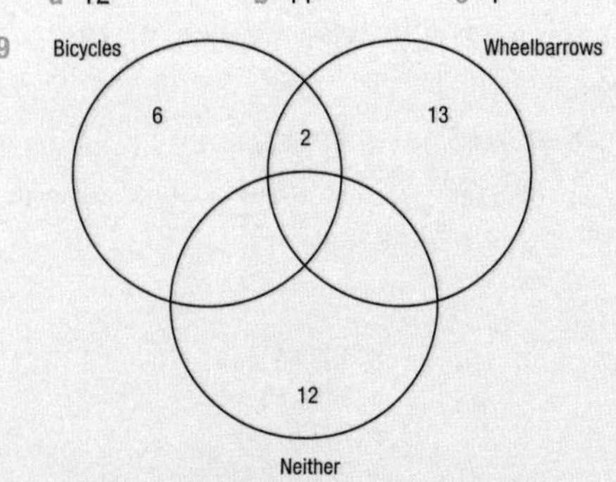

a 6 b 13 c 2 d 25

10 a–c Answers will vary.

11 a–c Answers will vary.

CHALLENGE:

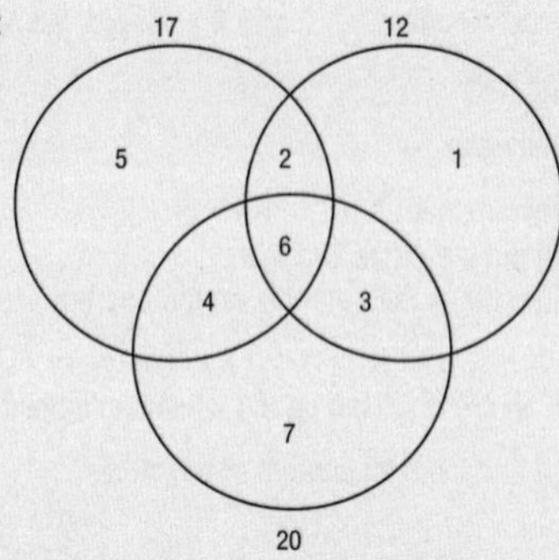

Labels and statements for this Venn diagram will vary.

Learning Unit 3: In the Tool Shed

Lesson 1

1–5 Answers will vary.

6 Students to make 2 tools.

Lesson 2

1 a 2 b 4 c A straight angle

2 a 60° b 50° c 47°
d 95° e 55° f 79°
g 90° h 75° i 15°

3 a–b Students to make the Egyptian tool.

4 Students to cut out a piece of cardboard.

CHALLENGE: Answers will vary (but the three numbers given should be multiples of 3:4:5)

5 Students to make a 45° set square.

6 Answers will vary.

7 Students to make a 30° and 60° set square.

Lesson 3

1 a 30° b 60° c 80°
d 85° e 45° f 70°

2 A = 30°, B = 70°, C = 15°

3 a 40° b 28° c 140° d 155°
e 39° f 109° g 132°

4 By finding the missing internal angle (60°) and then subtracting that answer from 180°. Hence $a = 120°$

5 a 140° b 132° c 157°
d 78° e 167° f 100°
g 120° h 132° i 142°

CHALLENGE: Students to perform an investigation.

Lesson 4

1 a Answers will vary; but students should have written two adjacent angles.

  b Answers will vary; but students should have written two angles that are not adjacent.

2 a 75°, Complementary b 27°, Complementary
  c 52°, Supplementary d 148°, Supplementary
  e 16°, Complementary f 34°, Complementary

3 a 85° b 71° c 53° d 34° e 16°

4 a 171° b 112° c 95° d 73° e 31°

5 Answers will vary.

6 Answers will vary.

7 a 122° b 40° c 80°
  d 30° e $y = 78°, x = 74°$

8 a–c Answers will vary.

9 a $y = 80°, x = 100°$
  b $y = 30°, x = 150°$
  c $a = 45°, b = 85°, c = 50°, d = 45°$
  d $p = 40°, q = 50°, r = 40°, s = 30°, t = 60°$

Lesson 5

1 a C b C c C d G e G f C

2 a D b A c B

3 a S b W c T

4 a–c Student's diagram may be different to this but must show a set of corresponding, alternate and co-interior angles.

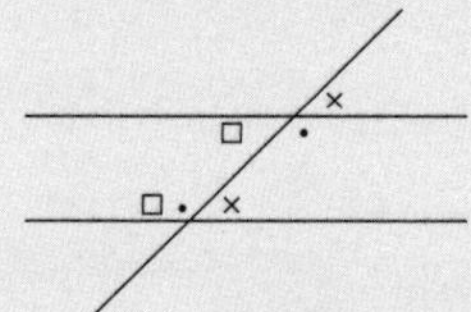

5 Explanations may vary.
  a 38° b 95° c 105° d 92° e 105°
  f 97° g 70° h 120° i 95° j 40°
  k $x = 54°, s = 126°, m = 126°$
  l $g = 83°, p = 138°, d = 83°$

CHALLENGE: Answers will vary.

Lesson 6

1

Polygon	Number of sides	Number of triangles within shape	Angle sum
	3	1	$1 \times 180° = 180°$
	4	2	$2 \times 180° = 360°$
	4	2	$2 \times 180° = 360°$
	5	3	$3 \times 180° = 540°$
	8	6	$6 \times 180° = 1080°$

2 a 540° b 720° c 360° d 720° e 1080° f 1080°

3 a 900° b 1260° c 1440°

4 a 150° b 130° c 130° d 120°
  e 113.33° f 99° g 115° h 110°

5 a 110° b 120° c $x = 95°, y = 85°$
  d 105° e 115° f 65°

6 a $y = 105°, x = 150°$ b $x = 130°, y = 102.5°$
  c $y = 110°, x = 110°$ d $x = 60°, y = 70°, z = 80°$
  e $y = 115°, x = 110°$ f $x = 100°, y = 80°, z = 100°$
  g $x = 60°, y = 120°$

CHALLENGE: Yes; it can.

Lesson 7

1 a $\frac{1}{2}$ inch b $\frac{5}{16}$ inch c $\frac{11}{32}$ inch
  d $\frac{5}{8}$ inch e $\frac{3}{4}$ inch f $1\frac{7}{8}$ inch

2 $\frac{1}{8}, \frac{1}{4}, \frac{5}{16}, \frac{11}{32}, \frac{1}{2}, \frac{9}{16}, \frac{5}{8}, \frac{11}{16}, \frac{3}{4}, 1\frac{13}{16}, 1\frac{7}{8}$

3 $\frac{7}{6} = 1\frac{1}{6}$ kg

4 a $\frac{103}{24} = 4.29$ L b 0.71 L c 1.38 L

5 a 18.67 m or $18\frac{2}{3}$ m b 10.5 m or $10\frac{1}{2}$ m
  c $\frac{7}{12} = 0.58$ m d 8.17 m or $8\frac{1}{6}$ m

CHALLENGE: $6 \times$ K2.20 = K13.2
Bundle of 6 = K11
Otto should buy the bundle of six for K11.

6 a 5 b 0.06 m or $\frac{1}{16}$ m

7 A and D
B and C
E and G and F
H and K
J and L

8 $4\frac{2}{3}$ L = 4.67 L

9 0.45 kg or $\frac{9}{20}$ kg

10 a $3\frac{1}{3}$ m b 4 m c $4\frac{1}{8}$ m d 2 m

11 a $3\frac{7}{24}$ m b $4\frac{3}{16}$ m c $4\frac{13}{40}$ m d $1\frac{29}{30}$ m

12 a 7 b $\frac{5}{8}$ c $\frac{9}{16}$

d Diagrams will vary (but should show the number of tools that are new/old and woodworking/non-woodworking).

13 a 39% b 18% c 35% d 30% e 75%
f 40% g 32% h 70% i 125% j $33\frac{1}{3}$%

14 a 72 L b 96 L c 120 L d 192 L e 300 L f 420 L

15 a K4.30 b K130.10 c K12.48 d K12.32
e K955.23 f K19.80 g K13.72 h K11

Lesson 8

1

Spanner sizes given as a fraction of an inch Decimal to fraction conversion chart								
Decimal fraction	0.25	0.313	0.375	0.445	0.5	0.625	0.75	0.875
Common fraction	$\frac{1}{4}$	$\frac{313}{1000}$	$\frac{3}{8}$	$\frac{89}{200}$	$\frac{1}{2}$	$\frac{5}{8}$	$\frac{3}{4}$	$\frac{7}{8}$

2 a $\frac{3}{5}$ b $\frac{1}{5}$ c $\frac{17}{20}$ d $\frac{9}{20}$ e $\frac{1}{20}$ f $\frac{3}{25}$
g $\frac{9}{50}$ h $\frac{23}{50}$ i $\frac{37}{100}$ j $\frac{93}{100}$ k $\frac{9}{25}$ l $\frac{6}{25}$

3

Spanner sizes given as a fraction of an inch Decimal to fraction conversion chart						
Decimal fraction	1.125	1.25	1.625	1.6875	1.875	2.050
Common fraction	$1\frac{1}{8}$	$1\frac{1}{4}$	$1\frac{5}{8}$	$1\frac{11}{16}$	$1\frac{7}{8}$	$2\frac{1}{20}$

4 a $1\frac{4}{5}$ b $2\frac{2}{5}$ c $6\frac{2}{25}$ d $3\frac{3}{4}$
e $8\frac{7}{25}$ f $14\frac{9}{25}$ g $4\frac{17}{200}$ h $2\frac{73}{500}$
i $5\frac{7}{200}$ j $6\frac{1}{8}$ k $23\frac{7}{250}$ l $10\frac{33}{125}$

5 a $\frac{3}{16}, \frac{1}{4}, \frac{197}{500}$; $\frac{197}{500}$ will handle the largest bolt
b $\frac{3}{5}, \frac{5}{16}, \frac{3}{8}$; $\frac{3}{5}$ will handle the largest bolt
c $\frac{63}{100}, \frac{3}{4}, \frac{11}{16}$; $\frac{3}{4}$ will handle the largest bolt
d $\frac{13}{16}, \frac{187}{250}, \frac{3}{8}$; $\frac{13}{16}$ will handle the largest bolt
e $\frac{41}{50}, \frac{7}{16}, \frac{1}{2}$; $\frac{41}{50}$ will handle the largest bolt

CHALLENGE: Answers will vary.

6 a 0.5 b 0.75 c 0.2 d 0.625 e 0.7 f 0.375
g 0.222 h 0.8 i 0.45 j 0.667 k 0.267 l 0.545

7 a 0.5, 0.3, 0.8 – $\frac{3}{10}, \frac{1}{2}, \frac{4}{5}$ b 0.6, 0.44, 0.57 – $\frac{4}{9}, \frac{4}{7}, \frac{3}{5}$
c 0.29, 0.45, 0.38 – $\frac{2}{7}, \frac{3}{8}, \frac{5}{11}$ d 0.3, 0.36, 0.38 – $\frac{3}{10}, \frac{4}{11}, \frac{5}{13}$
e 0.8, 0.88, 0.67 – $\frac{2}{3}, \frac{4}{5}, \frac{7}{8}$ f 0.43, 0.4, 0.25 – $\frac{1}{4}, \frac{2}{5}, \frac{3}{7}$
g 0.55, 0.44, 0.43 – $\frac{3}{7}, \frac{4}{9}, \frac{6}{11}$ h 0.63, 0.54, 0.78 – $\frac{7}{13}, \frac{5}{8}, \frac{25}{32}$

CHALLENGE: 3, 6, 7, 9

8 a–h Answers will vary.

9 a–h Answers will vary.

Lesson 9

1 a 8:16 = 1:2 b 2:3 c 2:8 = 1:4
d 15:25 = 3:5 e 4:7 f 200:400 = 1:2

2 a 4:7 b 4:11 c 7:4 d 7:11

3 a 10:4 = 5:2 b 4:10 = 2:5 c 4:14 = 2:7 d 10:14 = 5:7

4 3:10

5 5:8

6 6:1

7 a 1:4 b 1:5 c 1:5 d 7:2
e 1:3 f 5:2 g 1:2 h 7:10
i 7:3 j 1:3 k 1:50 l 1:21

8 a 5:14 b 1:4 c 19:21
d 7:16 e 17:20 f 41:12

9 a 1:25 b 3:50 c 2:15 d 1:9
e 1:16 f 1:8 g 7:3

10 a $\frac{1}{3}$ b $\frac{1}{5}$ c $\frac{2}{11}$ d $\frac{10}{7}$
e $\frac{1}{9}$ f 1 g $\frac{2}{5}$ h $\frac{7}{12}$
i $\frac{9}{28}$ j $\frac{1}{2}$ k $\frac{6}{47}$ l $\frac{2}{25}$

11 a 1:20 b 1:5 c 14:1 d 1:4
e 1:15 f 1:20 g 1:4 h 35:48
i 1:2 j 4:1 k 5:6 l 22:35

Lesson 10

1 1400 mL

2 660 g

3 15 mL

4 Powder = 1 L, Water = 800 mL

5 Screenings = 9, Sand = 6, Cement = 3

6 Bark = 10 kg, Leaves = 4 kg, Sawdust = 2 kg

CHALLENGE: Bark = 10.5 kg, Leaves = 4.5 kg

7 $\frac{3}{8}$ kg

8 $33\frac{1}{3}$%

9 $\frac{2}{3}$

10 20%

11 1.5 L

12

Percentage	Fraction	Decimal	Ratio
60%	$\frac{3}{5}$	0.6	3:5
50%	$\frac{1}{2}$	0.5	1:2
25%	$\frac{1}{4}$	0.25	1:4
75%	$\frac{3}{4}$	0.75	3:4
15%	$\frac{3}{20}$	0.15	3:20
33%	$\frac{1}{3}$	0.33	1:3

Learning Unit 4: Additional Learning, Revision and Assessment

Lesson 1

1

×	3	2	1	0	−1	−2	−3
3	9	6	3	0	−3	−6	−9
2	6	4	2	0	−2	−4	−6
1	3	2	1	0	−1	−2	−3
0	0	0	0	0	0	0	0
−1	−3	−2	−1	0	1	2	3
−2	−6	−4	−2	0	2	4	6
−3	−9	−6	−3	0	3	6	9

2 a positive b negative

3 a −28 b 40 c 12 d −30 e 80 f −63
g −24 h −24 i 126 j −40 k 360 l 560

4 a −7 b −5 c 8 d 2 e −3 f −12
g −4 h 5 i −3 j −40 k 2 l 1

5 + × + = + Two like symbols make a positive + ÷ + = +
− × − = + − ÷ − = +
+ × − = − Two unlike symbols make a negative + ÷ − = −
− × + = − − ÷ + = −

6 a −4 b 5 c 11 d −30 e −0.25
f 18 g −33 h 16 i 33

CHALLENGE: Answers will vary.

Lesson 2

1 a 500 seconds b K1.80 c 90t d 500 seconds

2 a $6n + 3$ b $8n + 1$ c $4n + 6$

3 a $14mn + 2n + 4m$ b $3 + 8x + 10xy$
c $6p + 6pn + n$ d $4xy + 1 + 3p$

4 a $10n + 24s$ b $6c + 12x - 12$ c $\frac{k}{5} + 6b$
d $12m + 9g$ e $3x + \frac{8n}{3}$ f $xy - (\frac{2x}{3y})$

5 a $4x - 4y$ b $9p + 3g$ c $30k + 12$
d $9m - 27p$ e $3yx - 6x$ f $5xy + 3y$

6 a $10 + 2m$ b $15m - 30m^2 + 4n$
c $35s - 9$ d $3 + 4h - hd - 4d$
e $21 + s$ f $-3p + 60$

7 a 6 b 48 c 16 d 0
e 20 f 80 g 480 h 160

8 a $3(p + 6g)$ b $6(2xy - xy)$ c $2(10hj + 7hj)$ d $3r(3p + 7)$
e $7w(2y + 1)$ f $3g(5k - 1)$ g $2q(9p + 1)$ h $12r(3p + 1)$

9 a 2 b −4 c −15 d −10
e −13 f −5 g 0 h 47

10 a 46 cm b 294 mm c 1.5 m d 3.9 m

11 a 125 mm b 13 cm c $2\frac{1}{8}$ m d 2.6 m

12 a 9.42 m b 31.4 cm c 125.6 mm d 15.7 m

13 a 31.4 mm b 157 cm c 157 cm d 6.28 m

14 Estimates will vary.

15 a 78.5 cm^2 b 314 mm^2 c 200.96 mm^2

16 a 88.31 cm^2 b 28.26 cm^2 c 63.59 cm^2

17 457 cm^2

18

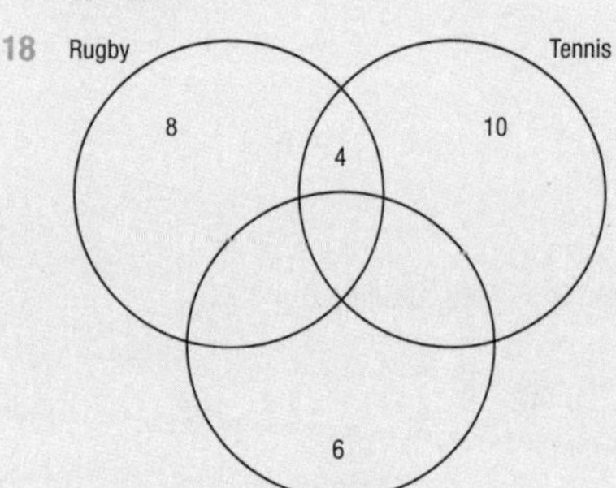

a 8 b 10 c 4 d 14

19 a 60° b 80° c 157° d 163°

20 Answers will vary.

21 a 162° b 135° c 111° d 22° e 5°

22 a 72° b 95° c $y = 80°, x = 65°$

23 a 42° b $x = 56°, y = 124°$ c $x = 145°, y = 35°$

24 a 540° b 720° c 900°

25 a $n = 70°, y = 100°, p = 80°, x = 80°, m = 115°$
b $x = 75°, y = 175°$
c $x = 65°, y = 34°, m = 130°, n = 82°$

26 a $4\frac{8}{21}$ b $2\frac{3}{20}$ c $5\frac{31}{40}$ d $2\frac{7}{24}$

27

Percentage	Decimal	Fraction	Ratio
26%	0.26	$\frac{13}{50}$	13:50
38%	0.38	$\frac{19}{50}$	19:50
70%	0.7	$\frac{7}{10}$	7:10
40%	0.4	$\frac{2}{5}$	2:5

28 a 17.5 b 18 c 45 d K0.52
e 63 f 0.3 m g 2.79 m

29 a–d Answers will vary.

30 a 457.14 mL
b 245 mL
c Water = 2 L, Chemical concentrate = 3.5 L

31 a 1:30 b 1:5 c 10:1 d 1:8 e 1:30 f 2:25

32 a −15 b 42 c 16 d −21

33 a −3 b −5 c −6 d 7

34 a −29 b 4 c 4

Assessment

Task 1

Answers will vary.

Task 2

1 K15

2 5 hours

3 33.33 minutes

4 1250 mL

5 Student A = K42, Student B = −K2, Student C = K11, Student D = −K10

6 a 5.47 b 42 c −54

7 a 44 cm
b 15 cm
c Answers will vary, but should be about 3 times the diameter.

8 Estimates will vary.

9 176.63 cm^2

10 Answers will vary

11 a 540° b 720°

12 a 152°, Supplementary b 50°, Complementary
c 80°, Supplementary d 75°, Complementary

13 a $b = 52°$, $d = 60°$, $a = 68°$, $c = 68°$
b $b = 24°$, $d = 20°$, $e = 100°$, $a = 36°$, $c = 36°$

14 a 300 km b $5\frac{1}{4}$ hours

15 a 12 hours b K78.75 c 21t

16 a

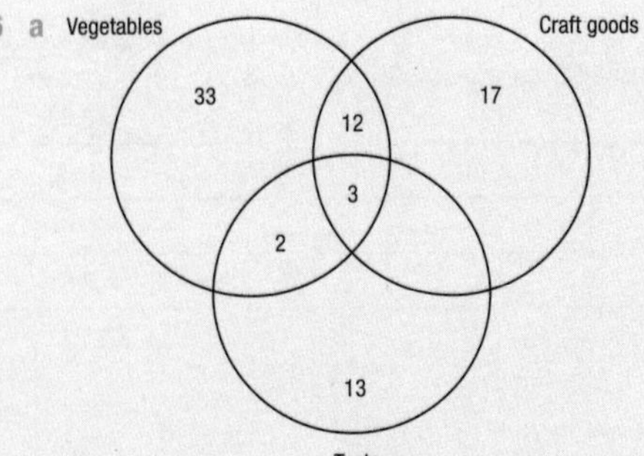

b 17 c 0 d 33

17 a 56 b 54 c 114

18 a 24 b 16 c −14 d 75 e 9

19 a $23 + 2t$ b $17m + 3$
c $24k - 10kh$ d $7h - hx - 4x$
e $4p - 10t + 12$ f $11k - 4k^2$

Topic 2

Learning Unit 1: The Games People Play

Lesson 1

1–7 Answers will vary.

Lesson 2

1 a 6:3 b 6:1 c 3:10 d 6:10 e 3:1

2 a A button; because there are more buttons in the bag than any other item.
b A seed; because there are less seeds in the bag than any other item.

3 a 5:1:1 – Buttons:Coins:Seeds
b A button
c 5:7

4 a 2:5 b 5:24 c 11:2 d 11:2 e 5:1 f 3:17

5 a 4:2 or 2:1 b 4 c 10
d A coloured card; because there are more of them in the deck.

6 Answers will vary.

CHALLENGE: Answers will vary.

7 a No; because there are five numbers that can be rolled for Player 2 to win and only one for Player 1 to win.
b Yes; because both players have an even chance of winning.
c No, because there are only two numbers above 4 and there are three numbers below it.
d Yes; because both players have an even chance of winning.
e Yes; because both players have an even chance of winning.
f Yes; because both players have an even chance of winning.

8 a 50; because both players have an even chance of winning, and 50 is half of 100.
b 75; because both players have an even chance of winning, and 75 is half of 150.
c 75; because both players have an even chance of winning, so if Sam has won 75 times then Rodney probably would have too.

9 a 33 or 34; because Jane has a one third chance of winning, and 33.3 recurring is one third of 100.
b 80; because Betty has a two thirds chance of winning and 80 is two thirds of 120.
c 80; because Betty has a 2:1 chance of winning and 80:40 = 2:1.

Lesson 3

1 a 50; because there is an equal chance of a card that is thrown into the air landing playing side up as landing playing side down, and 50 is half of 100.
b $\frac{50}{100} = 0.50$

2 a Answers will vary. b Answers will vary.

3 Answers will vary.

4 a 8 or 9; because one sixth of fifty is 8.3 recurring.
b $\frac{8}{50} = 0.16$ or $\frac{9}{50} = 0.18$

5 a Answers will vary. b Answers will vary.

6 Answers will vary.

7 a A = 0.225 B = 0.275 C = 0.2 D = 0.3

b 0.25; because each pattern has an equal chance of being spun; and the more times you spin the more likely the results will come out as predicted.

CHALLENGE: Answers will vary.

Lesson 4

1 Answers will vary.

2 a Certain b Likely or Unlikely
c Answers will vary. d Unlikely
e Impossible f Unlikely
g Answers will vary. h Certain
i Answers will vary. j Answers will vary.

3 a–e Answers will vary.

4 a $\frac{1}{2}$ b $\frac{1}{13}$ c Answers will vary.
d 0.5 e 25% f Answers will vary.

5 a Pr $\frac{13}{52}$ = Pr $\frac{1}{4}$
b Pr $\frac{4}{52}$ = Pr $\frac{1}{13}$
c Pr $\frac{26}{52}$ = Pr $\frac{1}{2}$
d Pr $\frac{36}{52}$ = Pr $\frac{9}{13}$ (Assuming an Ace is not a number card)
e Pr $\frac{1}{6}$
f Pr $\frac{5}{10}$ = Pr $\frac{1}{2}$
g Pr $\frac{3}{10}$

CHALLENGE: $\frac{5}{36}$

6 a Pr $\frac{1}{30}$ b Pr $\frac{9}{30}$ = Pr $\frac{3}{10}$
c Pr $\frac{3}{30}$ = Pr $\frac{1}{10}$ d Pr $\frac{5}{30}$ = Pr $\frac{1}{6}$

7 a Pr $\frac{2}{8}$ = Pr $\frac{1}{4}$ b Pr $\frac{1}{3}$ c Pr $\frac{4}{6}$ = Pr $\frac{2}{3}$ d Pr $\frac{3}{8}$

8 a–e Answers will vary.

CHALLENGE: Pr $\frac{1}{3}$ (B)

Lesson 5

1 a 180 b 3 c 67

2 Answers will vary.

3 a–d Answers will vary.

4 Answers for this question assume that all three darts hit the board. These answers show some ways the score can be achieved, but there are other options.
a Bullseye, bullseye, bullring / Triple 20, triple 20, 5 / Triple 20, double 20, bullring
b Bullseye, bullseye, bullseye / Triple 20, triple 20, double 15 / Triple 20, bullseye, double 20
c Triple 20, 5, 5 / Double 20, 20, 10 / Double 20, 15, 15
d 2, 2, 1 / 1, 1, 3 / Double 1, 1, 2
e 3, 3, 2 / 2, 2, 4 / 1, 5, 2
f 10, 1, 2 / 5, 5, 3 / 6, 6, 1
g Double 10, double 2, 1 / 12, 8, 5 / 7, 8, 10
h 1, 1, 1
i 7, 7, 7 / 5, 5, 11 / 19, 1, 1
j Double 10, double 1, 2 / 10, 10, 4 / 6, 8, 10

5 Answers will vary.

6 Answers will vary.

Lesson 6

1 Answers will vary.

2 a Multiplication
b Division or subtraction (depending on the circumstance)

3 Answers will vary.

4 Students to play a game.

5 18

6 Students to play a game.

7 Answers will vary.

8 Answers will vary.

CHALLENGE: 17 (and maybe some others that students come up with)

Lesson 7

1 a Pr $\frac{5}{200}$ = Pr $\frac{1}{40}$ b Pr $\frac{195}{200}$ = Pr $\frac{39}{40}$

2 a Pr $\frac{10}{10\,000}$ = Pr $\frac{1}{1000}$ b Pr $\frac{9990}{10\,000}$ = $\frac{999}{1000}$
c Pr $\frac{99}{100}$ d Pr $\frac{493}{500}$

3 a Pr $\frac{2}{6}$ = Pr $\frac{1}{3}$ b Pr $\frac{4}{6}$ = Pr $\frac{2}{3}$ c Pr $\frac{4}{6}$ = Pr $\frac{2}{3}$

CHALLENGE: K20

4 a

Sum of two dice	2	3	4	5	6	7	8	9	10	11	12
Combinations	1,1	1,2	1,3 2,2	1,4 2,3	1,5 2,4 3,3	1,6 2,5 3,4	2,6 3,5 4,4	3,6 4,5	4,6 5,5	5,6	6,6

b 6, 7 and 8

c Answers will vary, but should refer to numbers 6, 7 and 8 not being included in the winning numbers.

5 a It is not fair. Reasons will vary. b Answers will vary.

6 a Students to play a game. b Answers will vary.
c Answers will vary. d Answers will vary.

Lesson 8

1 a–c Answers will vary.

2 a–c Answers will vary.

3 a Answers will vary. b Answers will vary.

4 a K55 b Pr $\frac{1}{25}$ c Answers will vary.

CHALLENGE: $\frac{24}{25}$ (explanation will vary)

5 a Pr $\frac{1}{55}$ b Pr $\frac{2}{55}$ c Pr $\frac{2}{55}$ d Pr $\frac{3}{55}$

6 Answers will vary.

Lesson 9

1 a K17 b K3 c K6

2 No

3 Answers will vary.

4 $\frac{3}{17}$

5 K17

6 Answers will vary.

7 a National Capital District b Western Highlands
c National Capital District, Madang, Morobe

8 National Capital District; $Pr\ \frac{150}{27\,000} = Pr\ \frac{1}{180}$
Western Highlands; $Pr\ \frac{150}{6000} = Pr\ \frac{1}{40}$
Madang; $Pr\ \frac{150}{5200} = Pr\ \frac{3}{104}$
Morobe; $\frac{150}{11\,500} = Pr\ \frac{3}{230}$

9 Madang

10 Madang

11 a Answers will vary (but quite a low probability).
b Answers will vary (but quite a high probability).
c Answers will vary (but a high probability).
d Answers will vary (but quite a low probability).

12 a Answers will vary.
b Answers will vary (but more than $\frac{5}{8}$).

13 a Answers will vary. b Answers will vary.

Topic 2
Learning Unit 2: Water Sports

Lesson 1

1–8 Answers will vary.

Lesson 2

1 a $V = 120\text{ cm}^3$ b $V = 8\text{ m}^3$ c $V = 29.568\text{ m}^3$

2

	Dimensions			Volume (cm^3)	Capacity (mL)
	Length	Width	Height		
	3 cm	3 cm	3 cm	27 cm^3	27 mL
	12 cm	2 cm	2 cm	48 cm^3	48 mL
	10 cm	5 cm	4 cm	200 cm^3	200 mL
	15 cm	1 cm	0.5 cm	7.5 cm^3	7.5 mL
	4 cm	2 cm	12 cm	96 cm^3	96 mL
	8 cm	6 cm	5 cm	240 cm^3	240 mL

3 a $V = 3000\text{ cm}^3$ b $V = 7500\text{ cm}^3$ c $V = 64\text{ m}^3$ d $V = 48\text{ m}^3$

4 a 3000 mL b 7500 mL c 64 000 L d 48 000 L

5 a 50 mL b 85 mL c 1 250 000 L d 2500 L

6 a $V = 20\text{ cm}^3$ b $V = 330\text{ cm}^3$ c $V = 8000\text{ cm}^3$ d $V = 3250\text{ cm}^3$

7 a $V = 30\text{ cm}^3$ b $V = 250\text{ cm}^3$ c $V = 56\text{ cm}^3$ d $V = 375\text{ cm}^3$
e $V = 2000\text{ cm}^3$ f $V = 3750\text{ cm}^3$ g $V = 2300\text{ cm}^3$ h $V = 1780\text{ cm}^3$

8 a 15 mL b 85 mL c 180 mL d 675 mL
e 5 000 L f 2600 L g 3500 mL h 873 mL

CHALLENGE: 486 L

Lesson 3

1 a 76.2 cm^3 b 32.9 cm^3 c 144.64 cm^3 d 79.2 cm^3

2 a 58.88 L b 63.58 L c 4.5 L d 6.3 L

3 Answers will vary.

4 168 200 L

5 435 000 L

6 Answers will vary.

7 a 100.48 mL b 307.76 mL c 1.26 L
d 540 000 L e 4.71 L

CHALLENGE: 1589.63 L

Lesson 4

1 a 100 gallons b 100 L c 100 quarts
d 50 L e 100 fl. oz f 50 quarts

2 a 110 gallons b 2025 L c 528 quarts
d 500 L e 132 pints f 75 L
g 8 L h 875 fl. oz i 9 L

3 a 10 gallons b 66 quarts c 300 L

4 a 10 gallons, 500 fl. oz, 10 L, 4 quarts
b 2 gallons, 10 pints, 100 fl. oz, 1 L
c 2 gallons, 5 L, 3 quarts, 30 fl. oz
d 4 gallons, 5 quarts, 5 L, 100 fl. oz

5 All these answers should have working accompanying them.
a 1 litre of water b 8 pints of oil
c 10 gallons of petrol d 50 litres of milk

6 a 100 m b 100 cm c 1 km
d 2 yards e 10 miles f 50 inches

7 a 4 feet b 600 cm c 6 inches
d 1000 mm e 8 km f 20 miles
g 240 km h 4 feet i 6 m

8 a 1 yard, 0.3 m, 10 inches, 3 mm
b 0.5 km, 10 m, 10 yards, 300 cm
c 5 km, 1 mile, 20 yards, 5 inches
d 0.3 km, 50 yards, 5 m, 15 feet

9 a 11, 44, 20 b 44, 44, 20 c 1.76, 35, 200 fl. oz
d 250 mm e 4.8, 3, 1000

Lesson 5

1 All these answers should have working accompanying them.
a 61.56 b 70.47 c 50.4 d 62.25 e 67.2

2 72.21

3 58.32

4 0 points different

5 0.72 points different

6 a Answers will vary. b Answers will vary.

CHALLENGE: Answers will vary.

7 a 58.65 b 59.74 c 55.91

8 a All scores of 7 with a degree of difficulty at 3.
b 36 and 63

Lesson 6

Answers for this lesson are all rounded to 2 decimal places and each subsequent question uses the previously rounded number.

1 Answers will vary.

2 1.86 m/sec

3 5.37 sec

4 Answers will vary.

5 2.05 sec

6 1 min 58.36 sec

7 Answers will vary.

8 0.62 sec

9 1 min 58.56 sec

10 7 min 50.96 sec

11 2.55 sec

12 4.72 sec

13 6.07 sec

14 About 16.67 m (with explanation)

15 Answers will vary.

16 5.53 sec

17 a–d Answers will vary.

CHALLENGE: Answers will vary.

Lesson 7

1 432

2 a 369 b 19.25 L c 36 000

3 a 1.1 b 1.15 c 1.38
d 1.07 e 1.047 f 1.175

4 a 0.75 b 0.82 c 0.54
d 0.97 e 0.972 f 0.882

5 a 47.5 b 77.52 c 84.46 d 131.04

6 a 38.25 b 26.24 c 68.62 d 70.38

7 a 23 b 28.75 (rounded to 29)
c 51.75 (rounded to 52) d 36.8 (rounded to 37)

8 a 20% b 25% c 35%

9 a 25% b 20% c 200%
d 300% e 25% f 40%

10 a Didn't increase by more than 10%
b Did increase by more than 10%
c Didn't increase by more than 10%
d Did increase by more than 10%

CHALLENGE: (All numbers rounded down to the nearest whole person)

Year	Men	Women	Children
2004	500	400	200
2005	575	460	230
2006	661	529	264
2007	760	608	303

11 a 25% b 20% c 24%
d 25% e 25% f 10%

12 a Incorrect (K42.50) b Correct
c Incorrect (K10.20) d Incorrect (K15.30)

Lesson 8

1 a 4 b 16 c 6 d 2.7

2 a 3 b 8 c 0 d 1

3 Answers will vary.

4 Swimmer A = 28.62
Swimmer B = 28.49

5 a 6 b 10.5 c 8 d 4
e 11 f 6.5 g 2.6

6 Answers will vary (but the mean of the middle two numbers must equal 3.5).

7 a 2 b 4 and 11 c 15
d 3 e No mode f 2

8 a Answers will vary (but 7 must be the mode).
b Answers will vary (but there must be two modes).
c Answers will vary (but there must be no mode).

9

	Data Set	Mean	Median	Mode
a	23, 4, 7, 4, 9	9.4	7	4
b	2, 5, 15, 8, 11, 12	8.83	9.5	No Mode
c	1.3, 3.2, 1.03, 4, 1.3	2.17	1.3	1.3
d	10.3, 9.2, 9.01, 10.03	9.64	9.62	No Mode
e	11, 15, 11, 23, 9, 7, 9	12.14	11	9 and 11

10 a 69.5 b 72 c 72 d Answers will vary.

Lesson 9

1 a Answers will vary (but a pictogram must be drawn).
  b Answers will vary.

2 a Graph below or something to the same effect.

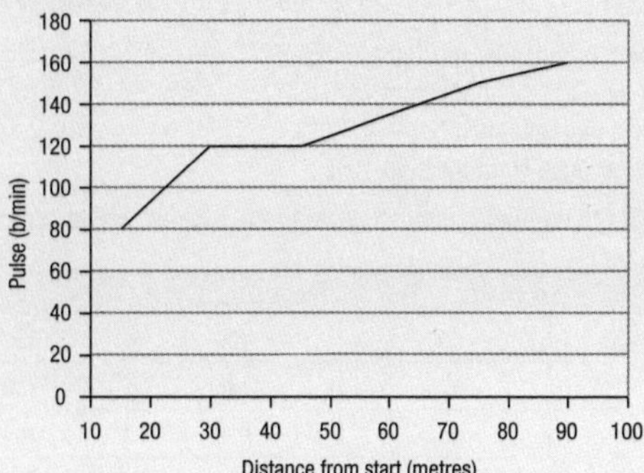

  b It rises
  c Answers will vary (but should be 160 or above).

3 a Chart below or something to the same effect.

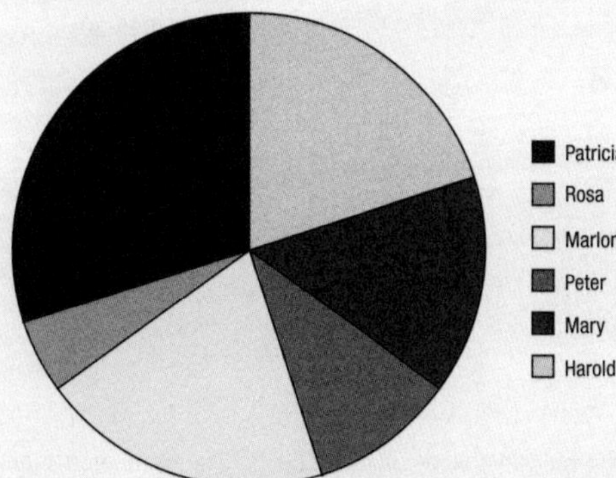

  b Patricia. $\frac{6}{20} = \frac{3}{10}$
  Rosa. $\frac{1}{20}$
  Marlon. $\frac{4}{20} = \frac{1}{5}$
  Peter. $\frac{2}{20} = \frac{1}{10}$
  Mary. $\frac{3}{20}$
  Harold $\frac{4}{20} = \frac{1}{5}$

4 a Diving score b Trial number c The 4th
  d 10 e 60–70

5 a Time in the water (in minutes)
  b A; because it is very unlikely that Peter would ever have a heart rate of zero.
  c B; because the difference between the top heart rate and bottom heart rate is greater in graph B than graph A.
  d A
  e There would not have been a big spike in his heart rate at the five minute mark.

6 a Answers will vary (but a graph should be drawn and labelled).
  b Answers will vary.

7 a 0 m
  b 20 m
  c Between the 6 and 8 minute mark
  d 10 minutes

8

Attempt	Time (in min)
1	2:43
2	2:42
3	2:39
4	2:40
5	2:36
6	2:33
7	2:33
8	2:30

Lesson 10

1 a Answers will vary, but should include the fact that the two y-axes have different scales.
  b That it was very low
  c Graph A: Reasons will vary.

2 a Yes
  b 53
  c June
  d That there was a big increase in clients over the six month period
  e It has been drawn so that the y-axis starts at 50 and increases by one client each time.
  f Answers will vary.

3 a Yes
  b The person who enjoys watching diving immensely, because they have made their graph so that it looks like lots of people attended the diving events.
  c The scale on the y-axis
  d 3 pm
  e 100

4 a 10
  b 2
  c The scale on the y-axis is misleading.
  d Answers will vary.
  e Answers will vary (but will include a graph and comment).

5 a 9 (assuming one picture represents one medal)
  b There is no indication of how many medals each picture represents and the pictures should be uniform sizes.
  c Answers will vary (but will include a pictogram).

6 Answers will vary.

Learning Unit 3: In the Bag

Lesson 1

1–3 Answers will vary.

Lesson 2

1

Solids with:		
all plane surfaces	all curved surfaces	a combination of plane and curved surfaces
A, C, D, H, J, K	E	B, F, G, I

2 a Answers will vary, but three non-rectangular prisms should be drawn.
b Answers will vary depending on a.
c Answers will vary depending on a.

3 a Answers will vary, but three non-square pyramids should be drawn.
b Answers will vary depending on a.
c Answers will vary depending on a.

CHALLENGE: Answers will vary (but a solid shape that is not a prism, pyramid, cylinder or cone should be drawn).

4 a Triangles, 4 b Squares, 6
c 2 circles, 1 rectangle d Rectangles, 6

5 a Cylinder b Cube and triangular prism
c Rectangular prism d Triangular prism
e Sphere f Cone
g Cylinder h Triangular prism

6 a–f Answers will vary.

Lesson 3

1 a Students to draw a net for a rectangular prism.
b Rectangle
c Yes; students to draw a diagram showing rectangular prisms stacked.

2 a Templates will vary.
b Templates will vary.
c Yes; students to draw a diagram showing hexagonal prisms stacked.

3 a Students to fold net into a solid.
b Dodecahedron
c They will stack but there will be spaces between the solids.

4 a–f Nets drawn by students will vary.

5 a  b 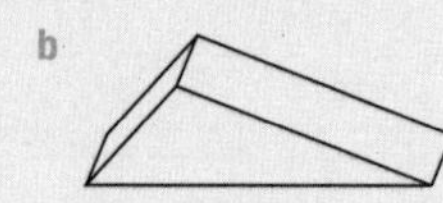

c 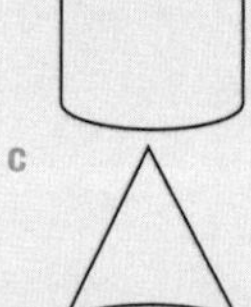 d 

6 Students to make an octahedron.

7 a and b

8 b, d, e

9 a, b, e

CHALLENGE: Answers will vary.

Lesson 4

1 a 4 cm^3 b 15 cm^3 c 16 cm^3 d 40 cm^3 e 28 cm^3

2 a 24 cm^3 b 35 cm^3 c 29 cm^3 d 38 cm^3 e 29 cm^3

3 a 2 cm^3 b 3600 cm^3 c 3750 cm^3 d 8000 cm^3
e 460 cm^3 f 52 500 cm^3 g 30 m^3 h 350 mm^3

4 a 6120 cm^3 b 574.75 cm^3 c 84 000 cm^3

CHALLENGE: 588 000 cm^3

5 a 330 cm^3 b 192 cm^3 c 540 cm^3
d 512 cm^3 e 920 cm^3 f 410 cm^3

6 a 264 cm^3 b 360 cm^3 c 360 cm^3

7 a 300 cm^3 b 288 cm^3 c 750 cm^3
d 520 cm^3 e 1000 cm^3 f 84 m^3

8 a–d Answers will vary.

9 a 942 cm^3 b 6280 cm^3 c 1 570 000 cm^3 d 15 072 mm^3
e 169.56 cm^3 f 2512 m^3 g 10 597.5 cm^3 h 4521.6 m^3

Lesson 5

1 Answers will vary.

2 Answers will vary; but should mention that the larger the area of the base and the longer the height, the larger the volume will be.

3 Answers will vary.

4 Answers will vary; but should mention that the larger the area of the base and the longer the height, the larger the volume will be.

5 a 84 cm^3 b 85.32 cm^3 c 83.32 cm^3 d 160 cm^3
e 190 cm^3 f 96 cm^3 g 56 cm^3 h 70 cm^3

CHALLENGE: 1.69 cm (Depending on how much rounding occurs and when that rounding occurs, the answer could be somewhat different from this. Anywhere between 1.2 and 2.2 could be correct.)

Lesson 6

1 a 216 cm^3 and 282.6 cm^3
b 216 mL and 282.6 mL
c The plastic-coated cardboard prism
d Answers will vary, but there must be two drawings of shelves. One which has five boxes side by side, six deep and one which has five cans side by side, three deep.
e The plastic-coated cardboard prism with reasons such as: They will make more money per drink sold and they can store more of them in the space they have.

2 a 4200 cm^3
b 22 750 cm^3
c 90 cm^3
d 1055.04 cm^3
e 65.42 cm^3 (measuring only the cone)
f 31 500 cm^3

3 a B

4 a

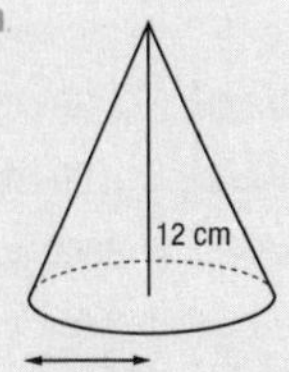

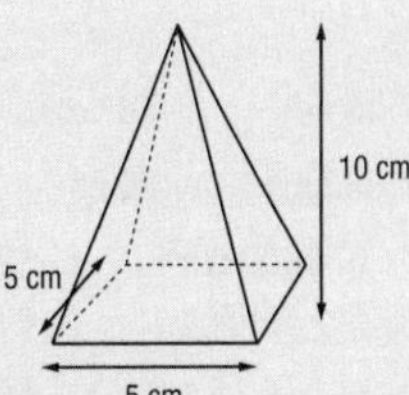

b Cone: $V = 200.96$ cm^3
Square Pyramid: $V = 83.32$ cm^3
The Cone container is larger.

5 Answers will vary.

6 Answers will vary.

7 Answers will vary.

8 a $x = 5$ cm b $x = 4$ cm c $x = 264$ cm^3
d $x = 3$ cm e $x = 47$ cm^3 f $x = 3$ cm

CHALLENGE: 2.32 cm

Lesson 7

1 a 1155.52 cm^3
b Answers will vary.
c Answers will vary depending on answer b.
d Answers will vary depending on answer c.
e Answers will vary depending on answer c.

2 a 125 cm^3
b Answers will vary.
c Answers will vary depending on answer b.
d Answers will vary depending on answer c.
e Answers will vary depending on answer c.

3 a Square Pyramid
b 4860 cm^3
c Answers will vary.
d Answers will vary depending on answer c.
e Answers will vary depending on answer c.
f Answers will vary depending on answer c.

4 Answers will vary.

5 Answers will vary.

6 Answers will vary.

7 a 25 b 12 m^2 (120, 000 cm^2)
c

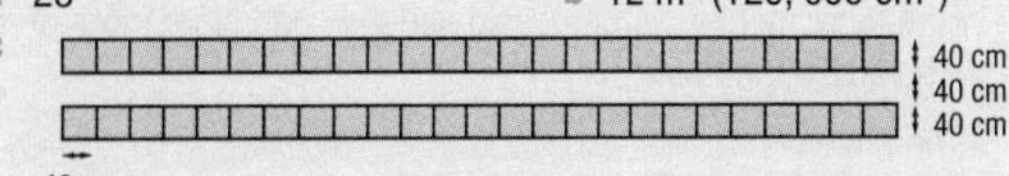

CHALLENGE: The single row requires a lesser area, but the three rows accommodate more spectators.

8 a 192
b 576 000 cm^2 (57.6 m^2)
c Answers will vary.
d Answers will vary; depending on the answer to c.

9 Answers will vary.

Lesson 8

1

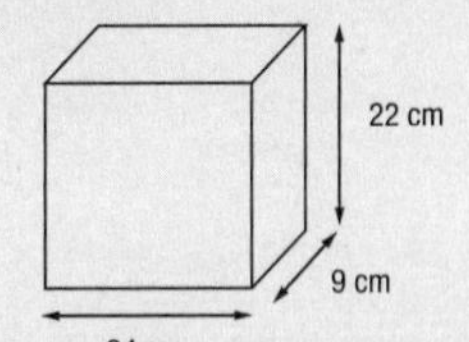

2 a 15 cm b 21 cm c 9 m d 12.38 mm
e 13.13 m f 7.25 km g 11.81 cm h 8.53 m

3 a 61 b 5 cm

CHALLENGE: $\frac{49}{244}$

4 6.875 hrs or $6\frac{7}{8}$ hrs

5 734

6 a $<$ b $<$ c $<$ d $>$
e $<$ f $<$ g $<$ h $<$

7 a 10
b 1.17 L
c 9
d They will all decrease. (a 9, b 0.875 L, c 8)

8 a $\frac{26}{21} = 1\frac{5}{21}$ b $\frac{51}{40} = 1\frac{11}{40}$ c $\frac{1}{5}$ d $\frac{2}{9}$
e $\frac{27}{80}$ f $\frac{13}{150}$ g $\frac{3}{2} = 1\frac{1}{2}$ h $\frac{32}{11} = 2\frac{10}{11}$

9 a Day 1: $\frac{7}{16}$, Day 2: $\frac{1}{2}$, Day 3: $\frac{9}{16}$
b $\frac{7}{16}$
c 52 500 L

CHALLENGE: 10 000 nights

10 a 4 laps b $\frac{3}{4}$ c 400 m

11 a 16 b 40 c 48 d 48

12 16

13 23

Lesson 9

1

Division algorithm	Fraction	Convert to decimal denominator	Decimal
$3 \div 4$	$\frac{3}{4}$	$\frac{75}{100}$	0.75
$1 \div 2$	$\frac{1}{2}$	$\frac{5}{10}$	0.5
$7 \div 10$	$\frac{7}{10}$	$\frac{7}{10}$	0.7
$4 \div 5$	$\frac{4}{5}$	$\frac{8}{10}$	0.8
$21 \div 100$	$\frac{21}{100}$	$\frac{21}{100}$	0.21
$13 \div 20$	$\frac{13}{20}$	$\frac{65}{100}$	0.65
$3 \div 8$	$\frac{3}{8}$	$\frac{375}{1000}$	0.375
$1 \div 3$	$\frac{1}{3}$	$\frac{33}{100}$	0.33

2 a Answers will vary. b Answers will vary.

3 a $12.75 = 12\frac{3}{4}$ b $5.8 = 5\frac{8}{10} = 5\frac{4}{5}$ c $3.05 = 3\frac{1}{20}$
d $1.8 = 1\frac{8}{10} = 1\frac{4}{5}$ e $35.91 = 35\frac{91}{100}$ f $9.74 = 9\frac{93}{125}$
g $6.5 = 6\frac{1}{2}$ h $13.5 = 13\frac{1}{2}$ i 5
j $26.67 = 26\frac{67}{100}$

4 a–j Answers the same as 3; but working will show all numbers as decimals before final answer rather than common fractions.

5 a K7.80 b K39.20 c K8.75 d K5.06

6 K44.22

7 a 26.2 L b 30.6 L c 44.2 L d 54.4 L

8 a 20 b 9 c 17 d 43 e 39 f 30

CHALLENGE: Answers will vary.

9 a $\frac{1}{8}$ b $\frac{3}{4}$ c $\frac{1}{20}$ d $\frac{1}{6}$ e $\frac{1}{4}$ f $\frac{1}{36}$

10 a 15.6 cm^3 b 20.8 cm^3 c 7.28 cm^3 d 51.2 cm^3
e 113.6 mm^3 f 37. 87 mm^3 g 5.76 cm^3 h 11.52 cm^3

Learning Unit 4: Additional Learning, Revision and Assessment

Lesson 1

1

Sequence in pattern	1st	2nd	3rd	4th	5th	6th	7th	8th	9th	10th
Calculation	1 × 1	2 × 2	3 × 3	4 × 4	5 × 5	6 × 6	7 × 7	8 × 8	9 × 9	10 × 10
Product	1	4	9	16	25	36	49	64	81	100

2 a the 15th square number is 15 × 15 = 225
b the 20th square number is 20 × 20 = 400
c the 40th square number is 40 × 40 = 1600
d the 100th square number is 100 × 100 = 10 000
e the 200th square number is 200 × 200 = 40 000

3 a 625 b 10 c 41 d 7 e 25
f 83 g 25 h 622 i −19 j 20

4 a Students to work with a partner (but end answer is Yes).
b Yes, yes, yes
c Answers will vary.

5 a

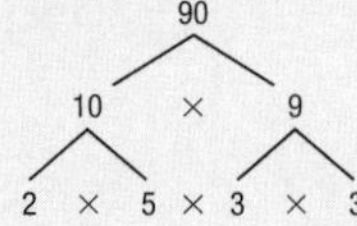

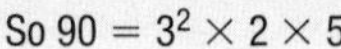

So $90 = 3^2 \times 2 \times 5$

b

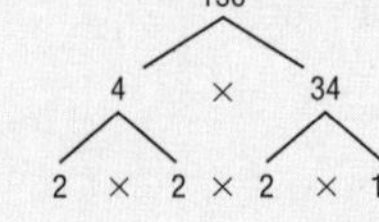

So $136 = 2^3 \times 17$

c

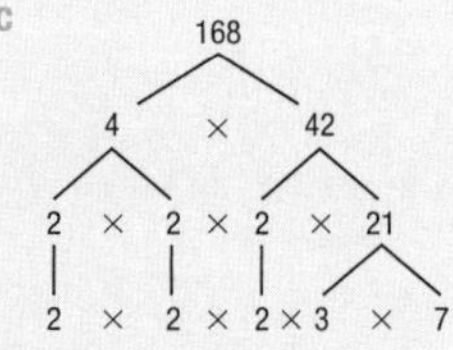
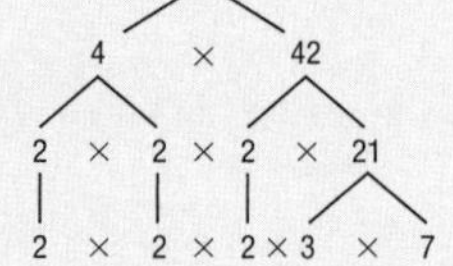

So $168 = 2^3 \times 3 \times 7$

6 a

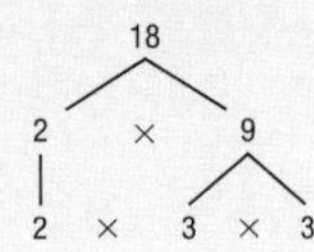

So $18 = 3^2 \times 2$

b

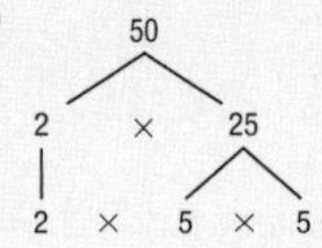

So $50 = 5^2 \times 2$

c

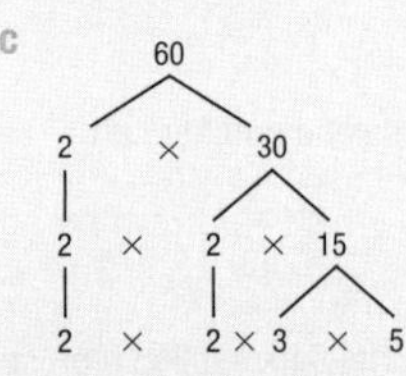

So $60 = 2^2 \times 3 \times 5$

d

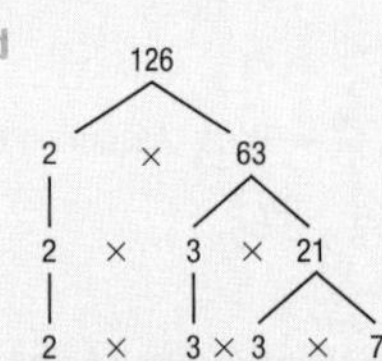

So $126 = 3^2 \times 2 \times 7$

e

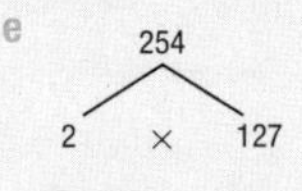

So $254 = 2 \times 127$

f

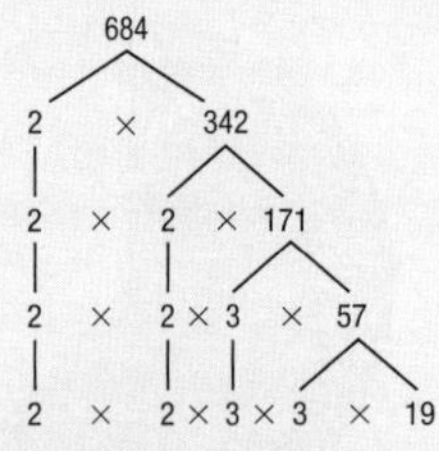

So $684 = 2^2 \times 3^2 \times 19$

g

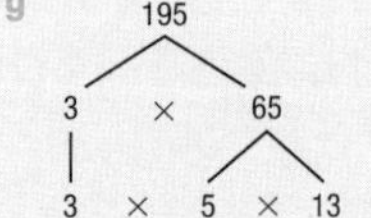

So $195 = 3 \times 5 \times 13$

h

156
2 × 78
2 × 2 × 39
2 × 2 × 3 × 13

So $156 = 2^2 \times 3 \times 13$

7 a $42 = 7 \times 3 \times 2$ b $30 = 2 \times 3 \times 5$
c $24 = 2^3 \times 3$ d $45 = 3^2 \times 5$
e $63 = 3^2 \times 7$ f $125 = 5^3$

8

Number	Prime factors	Index notation
225	3 × 3 × 5 × 5	$3^2 \times 5^2$
144	2 × 2 × 6 × 6	$2^2 \times 6^2$
216	2 × 2 × 2 × 3 × 3 × 3	$2^3 \times 3^3$
1800	2 × 10 × 10 × 3 × 3	$2 \times 10^2 \times 3^2$
2600	2 × 2 × 2 × 5 × 5 × 13	$2^3 \times 5^2 \times 13$

CHALLENGE: $2^3 \times 5^4 \times 830\,849$

Lesson 2

1 a 3:10 b 1:5 c 11:2
d 5:2 e 1:7 f 3:7

2 $\frac{40}{80} = \frac{1}{2}$

3 Pr $\frac{1}{6}$

4 $\frac{25}{100} = \frac{1}{4}$
$\frac{15}{100} = \frac{3}{20}$
$\frac{55}{100} = \frac{11}{20}$
$\frac{5}{100} = \frac{1}{20}$

5 a $Pr\frac{4}{52} = Pr\frac{1}{13}$
b $Pr\frac{26}{52} = Pr\frac{1}{2}$
c $Pr\frac{36}{52} = Pr\frac{9}{13}$ (Assuming that an Ace is not a number card)
d $Pr\frac{3}{6} = Pr\frac{1}{2}$
e $Pr\frac{1}{10}$

6 (Spinners below show possible answers. The dark section represents red.)

a

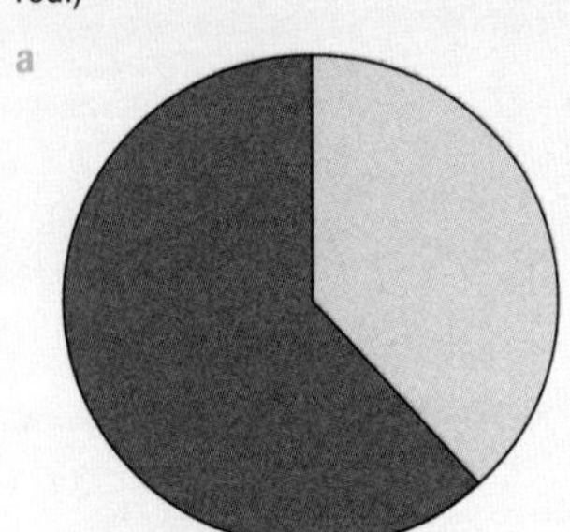

b

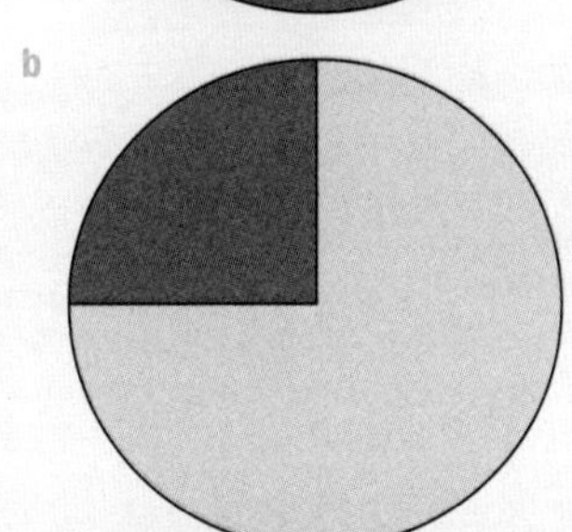

c

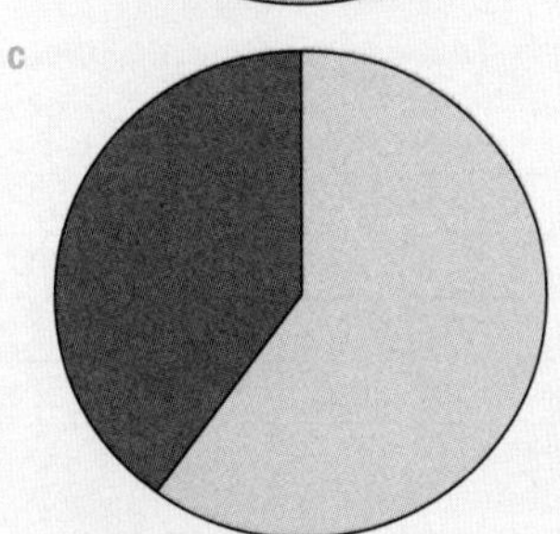

7 a $Pr\frac{10}{500} = Pr\frac{1}{50}$ b $\frac{Pr\ 490}{500} = Pr\frac{49}{50}$

8 a $Pr\frac{10}{200} = Pr\frac{1}{20}$ b $Pr\frac{190}{200} = Pr\frac{19}{20}$
c $Pr\frac{50}{200} = Pr\frac{1}{4}$ d K75
e $Pr\frac{150}{200} = Pr\frac{3}{4}$ f K100

9 a Team B b Team B c Team C

10 a 27 m^3 b 480 cm^3 c 100 m^3 d 2160 cm^3

11 a 20 mL b 35 mL c 1 300 000 L d 1750 L

12 76 800 L

13 a 22.5 L b 22 quarts c 5.5 gallons

14 10 miles, 1 km, 100m, 100 feet, 10 yards, 2 yards, 50 in, 1 m, 10 inches, 250 mm, 10 cm

15 a 30 b 97.92 c 54.59 d 147.42

16 a 55.25 b 45.92 c 78.96 d 74.52

17 a 87.5% b 83.32% c 700% d 100%
e 266.67% f 20%

18 a K34 b K12.75 c K21.25 d K15.30

19 a Mean = 11.32, Median = 11, Mode = 3
b Mean = 9.25, Median = 6.5, Mode = 16
c Mean = 3.48, Median = 2.6, Mode = No mode

20 a Percentage correct b Exam number
c 40 d Between 40 and 80

21 a Answers will vary; but should mention that one gives the impression that there is a big range and one gives the opposite impression.
b The scales on the y-axis are different.
c Answers will vary.
d Answers will vary.

22 Answers will vary (but student should draw three prisms).

23 Answers will vary (but student should draw three pyramids).

24 a, d

25 a 1728 cm^3 b 264 cm^3 c 360 cm^3
d 1130.4 cm^3 e 120 cm^3 f 45.3 cm^3

26 a $x = 4$ cm b $x = 2.23$ cm c $x = 15$ cm

27 a Answers will vary, but the volume of the box described should equal 6912 cm^3.
b Answers will vary depending on a.

28 a 15 b $\frac{1}{9}$ c 18

29 The answers for questions a–j should have working with both numbers converted to either fractions or mixed numbers.

a 3 b $6.4 = 6\frac{2}{5}$ c $2.3 = 2\frac{3}{10}$
d $0.85 = \frac{17}{20}$ e $5.18 = 5\frac{9}{50}$ f $5.075 = 5\frac{3}{40}$
g $10.24 = 10\frac{6}{25}$ h $8.75 = 8\frac{3}{4}$ i $2.57 = 2\frac{4}{7}$
j $8.88 = 8\frac{22}{25}$

30 a $\frac{5}{8}$ b $\frac{3}{28}$ c $\frac{7}{18}$

31 a 400 b 8 c 34 d 6 e 5

32 a

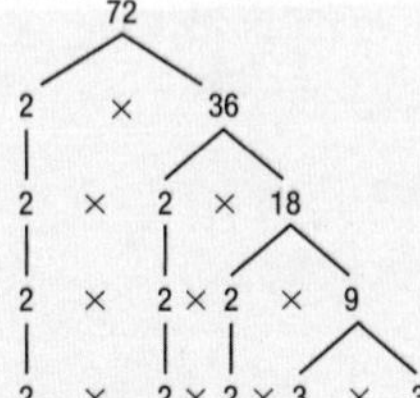

So $72 = 2^3 \times 3^2$

b

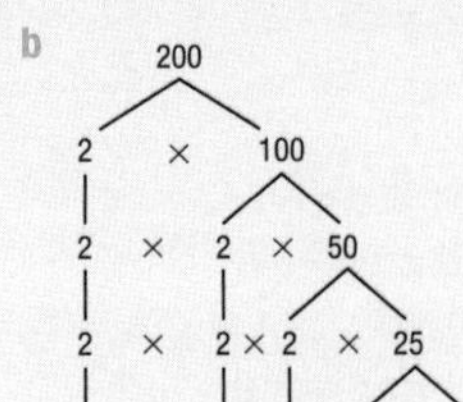

So $200 = 2^3 \times 5^2$

c

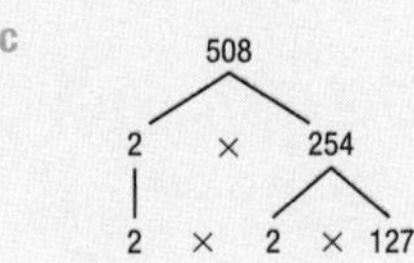

So $508 = 2^2 \times 127$

d

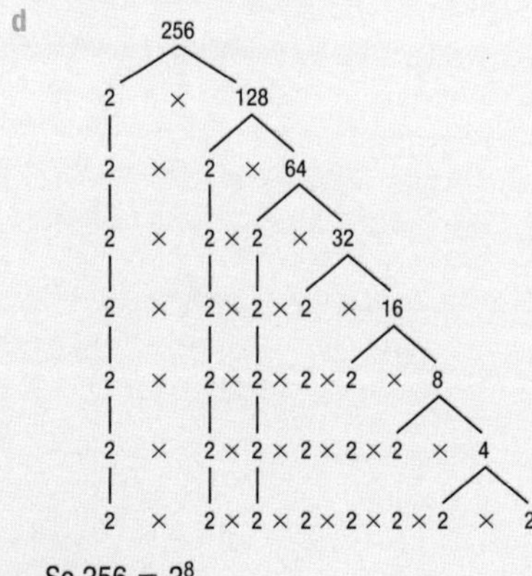

So $256 = 2^8$

Assessment

Task 1

Answers will vary (to be assessed by teacher).

Task 2

1 a 6:7 b 7:12 c 11:5 d 15:1

2 a 12 b 16

3 a 4:5 b 11:20 c 5:11 d 4:1

4 a $Pr\frac{10}{50} = Pr\frac{1}{5}$ b $Pr\frac{35}{50} = Pr\frac{7}{10}$

5 a $Pr\frac{2}{50} = Pr\frac{1}{25}$ b $Pr\frac{10}{50} = Pr\frac{1}{5}$

6 a 144 cm^3 b 904.32 cm^3

7 a 32 cm^3 b 261.67 cm^3

8 a 24 mL b 157 mL c 60 mL

9 6750 L

10 a 10 gallons, 5 L, 3 quarts, 100 fl. oz
b 2 gallons, 5 quarts, 1 L, 30 fl. oz
c 2 gallons, 10 pints, 5 L, 4 quarts
d 4 gallons, 500 fl. oz, 10 L, 100 fl. oz

11 a 5 km, 10 yards, 15 feet, 0.3 m, 20 cm
b 2 miles, 0.5 km, 300 cm, 1 yard, 12 inches
c 1 mile, 0.3 km, 20 yards, 10 m, 3 mm
d 3 km, 50 yards, 5 m, 200 cm, 10 inches

12 a 13.06 s b 0.79 s

13 a 32.59 s b 0.65 s

14 a 18.75% b $33\frac{1}{3}$% c 20% d 17.65%

15 a 70 b 70

16 a Yes; because the graph starts 0 km from home.
b 2 times
c Between 40 and 50 minutes; because that is where the most distance is achieved in a 10 minute period.

17 a Answers will vary (but student should draw a net for a prism).
b Answers will vary (but student should draw a net for a triangular pyramid).
c Answers will vary (but student should draw a net for a tapered solid other than a pyramid).

18 14

Notes

Notes

Acknowledgments

The authors and the publisher wish to thank the following copyright holders for granting permission to reproduce their material.

AAP Image/EPA/Patrick B. Kraemer, pp. 110, 117; Getty Images/Greg Wood, p. 112; iStockphoto/Tim McCaig, p. 32; Photolibrary/Alamy/picturesbyrob, p. 133; Irene Sawczak, pp. 2, 4, 6, 24, 28, 42, 43, 63, 95, 104, 108, 119, 120, 124, 126, 134, 145, 151; Shutterstock, p. 29.

Every effort has been made to trace the original sources contained in this book. The publisher would be pleased to hear from copyright holders to rectify any errors or omissions.